Pitman Research Notes in Mathematics Series

Submission of proposals for consideration
Suggestions for publication, in the form of outlines and representative samples, are invited by the Editorial Board for assessment. Intending authors should approach one of the main editors or another member of the Editorial Board, citing the relevant AMS subject classifications. Alternatively, outlines may be sent directly to the publisher's offices. Refereeing is by members of the board and other mathematical authorities in the topic concerned, throughout the world.

Preparation of accepted manuscripts
On acceptance of a proposal, the publisher will supply full instructions for the preparation of manuscripts in a form suitable for direct photo-lithographic reproduction. Specially printed grid sheets are provided and a contribution is offered by the publisher towards the cost of typing. Word processor output, subject to the publisher's approval, is also acceptable.

Illustrations should be prepared by the authors, ready for direct reproduction without further improvement. The use of hand-drawn symbols should be avoided wherever possible, in order to maintain maximum clarity of the text.

The publisher will be pleased to give any guidance necessary during the preparation of a typescript, and will be happy to answer any queries.

Important note
In order to avoid later retyping, intending authors are strongly urged not to begin final preparation of a typescript before receiving the publisher's guidelines and special paper. In this way it is hoped to preserve the uniform appearance of the series.

Longman Scientific & Technical
Longman House
Burnt Mill
Harlow, Essex, UK
(tel (0279) 426721)

Titles in this series

1 Improperly posed boundary value problems
A Carasso and A P Stone
2 Lie algebras generated by finite dimensional ideals
I N Stewart
3 Bifurcation problems in nonlinear elasticity
R W Dickey
4 Partial differential equations in the complex domain
D L Colton
5 Quasilinear hyperbolic systems and waves
A Jeffrey
6 Solution of boundary value problems by the method of integral operators
D L Colton
7 Taylor expansions and catastrophes
T Poston and I N Stewart
8 Function theoretic methods in differential equations
R P Gilbert and R J Weinacht
9 Differential topology with a view to applications
D R J Chillingworth
10 Characteristic classes of foliations
H V Pittie
11 Stochastic integration and generalized martingales
A U Kussmaul
12 Zeta-functions: An introduction to algebraic geometry
A D Thomas
13 Explicit *a priori* inequalities with applications to boundary value problems
V G Sigillito
14 Nonlinear diffusion
W E Fitzgibbon III and H F Walker
15 Unsolved problems concerning lattice points
J Hammer
16 Edge-colourings of graphs
S Fiorini and R J Wilson
17 Nonlinear analysis and mechanics: Heriot-Watt Symposium Volume I
R J Knops
18 Actions of fine abelian groups
C Kosniowski
19 Closed graph theorems and webbed spaces
M De Wilde
20 Singular perturbation techniques applied to integro-differential equations
H Grabmüller
21 Retarded functional differential equations: A global point of view
S E A Mohammed
22 Multiparameter spectral theory in Hilbert space
B D Sleeman
24 Mathematical modelling techniques
R Aris
25 Singular points of smooth mappings
C G Gibson
26 Nonlinear evolution equations solvable by the spectral transform
F Calogero
27 Nonlinear analysis and mechanics: Heriot-Watt Symposium Volume II
R J Knops
28 Constructive functional analysis
D S Bridges
29 Elongational flows: Aspects of the behaviou of model elasticoviscous fluids
C J S Petrie
30 Nonlinear analysis and mechanics: Heriot-Watt Symposium Volume III
R J Knops
31 Fractional calculus and integral transforms c generalized functions
A C McBride
32 Complex manifold techniques in theoretical physics
D E Lerner and P D Sommers
33 Hilbert's third problem: scissors congruence
C-H Sah
34 Graph theory and combinatorics
R J Wilson
35 The Tricomi equation with applications to the theory of plane transonic flow
A R Manwell
36 Abstract differential equations
S D Zaidman
37 Advances in twistor theory
L P Hughston and R S Ward
38 Operator theory and functional analysis
I Erdelyi
39 Nonlinear analysis and mechanics: Heriot-Watt Symposium Volume IV
R J Knops
40 Singular systems of differential equations
S L Campbell
41 N-dimensional crystallography
R L E Schwarzenberger
42 Nonlinear partial differential equations in physical problems
D Graffi
43 Shifts and periodicity for right invertible operators
D Przeworska-Rolewicz
44 Rings with chain conditions
A W Chatters and C R Hajarnavis
45 Moduli, deformations and classifications of compact complex manifolds
D Sundararaman
46 Nonlinear problems of analysis in geometry and mechanics
M Atteia, D Bancel and I Gumowski
47 Algorithmic methods in optimal control
W A Gruver and E Sachs
48 Abstract Cauchy problems and functional differential equations
F Kappel and W Schappacher
49 Sequence spaces
W H Ruckle
50 Recent contributions to nonlinear partial differential equations
H Berestycki and H Brezis
51 Subnormal operators
J B Conway

52 Wave propagation in viscoelastic media
F Mainardi
53 Nonlinear partial differential equations and their applications: Collège de France Seminar. Volume I
H Brezis and J L Lions
54 Geometry of Coxeter groups
H Hiller
55 Cusps of Gauss mappings
T Banchoff, T Gaffney and C McCrory
56 An approach to algebraic K-theory
A J Berrick
57 Convex analysis and optimization
J-P Aubin and R B Vintner
58 Convex analysis with applications in the differentiation of convex functions
J R Giles
59 Weak and variational methods for moving boundary problems
C M Elliott and J R Ockendon
60 Nonlinear partial differential equations and their applications: Collège de France Seminar. Volume II
H Brezis and J L Lions
61 Singular systems of differential equations II
S L Campbell
62 Rates of convergence in the central limit theorem
Peter Hall
63 Solution of differential equations by means of one-parameter groups
J M Hill
64 Hankel operators on Hilbert space
S C Power
65 Schrödinger-type operators with continuous spectra
M S P Eastham and H Kalf
66 Recent applications of generalized inverses
S L Campbell
67 Riesz and Fredholm theory in Banach algebra
B A Barnes, G J Murphy, M R F Smyth and T T West
68 Evolution equations and their applications
F Kappel and W Schappacher
69 Generalized solutions of Hamilton-Jacobi equations
P L Lions
70 Nonlinear partial differential equations and their applications: Collège de France Seminar. Volume III
H Brezis and J L Lions
71 Spectral theory and wave operators for the Schrödinger equation
A M Berthier
72 Approximation of Hilbert space operators I
D A Herrero
73 Vector valued Nevanlinna Theory
H J W Ziegler
74 Instability, nonexistence and weighted energy methods in fluid dynamics and related theories
B Straughan
75 Local bifurcation and symmetry
A Vanderbauwhede
76 Clifford analysis
F Brackx, R Delanghe and F Sommen
77 Nonlinear equivalence, reduction of PDEs to ODEs and fast convergent numerical methods
E E Rosinger
78 Free boundary problems, theory and applications. Volume I
A Fasano and M Primicerio
79 Free boundary problems, theory and applications. Volume II
A Fasano and M Primicerio
80 Symplectic geometry
A Crumeyrolle and J Grifone
81 An algorithmic analysis of a communication model with retransmission of flawed messages
D M Lucantoni
82 Geometric games and their applications
W H Ruckle
83 Additive groups of rings
S Feigelstock
84 Nonlinear partial differential equations and their applications: Collège de France Seminar. Volume IV
H Brezis and J L Lions
85 Multiplicative functionals on topological algebras
T Husain
86 Hamilton-Jacobi equations in Hilbert spaces
V Barbu and G Da Prato
87 Harmonic maps with symmetry, harmonic morphisms and deformations of metrics
P Baird
88 Similarity solutions of nonlinear partial differential equations
L Dresner
89 Contributions to nonlinear partial differential equations
C Bardos, A Damlamian, J I Díaz and J Hernández
90 Banach and Hilbert spaces of vector-valued functions
J Burbea and P Masani
91 Control and observation of neutral systems
D Salamon
92 Banach bundles, Banach modules and automorphisms of C*-algebras
M J Dupré and R M Gillette
93 Nonlinear partial differential equations and their applications: Collège de France Seminar. Volume V
H Brezis and J L Lions
94 Computer algebra in applied mathematics: an introduction to MACSYMA
R H Rand
95 Advances in nonlinear waves. Volume I
L Debnath
96 FC-groups
M J Tomkinson
97 Topics in relaxation and ellipsoidal methods
M Akgül
98 Analogue of the group algebra for topological semigroups
H Dzinotyiweyi
99 Stochastic functional differential equations
S E A Mohammed

100 Optimal control of variational inequalities
V Barbu
101 Partial differential equations and dynamical systems
W E Fitzgibbon III
102 Approximation of Hilbert space operators. Volume II
C Apostol, L A Fialkow, D A Herrero and D Voiculescu
103 Nondiscrete induction and iterative processes
V Ptak and F-A Potra
104 Analytic functions – growth aspects
O P Juneja and G P Kapoor
105 Theory of Tikhonov regularization for Fredholm equations of the first kind
C W Groetsch
106 Nonlinear partial differential equations and free boundaries. Volume I
J I Díaz
107 Tight and taut immersions of manifolds
T E Cecil and P J Ryan
108 A layering method for viscous, incompressible L_p flows occupying R^n
A Douglis and E B Fabes
109 Nonlinear partial differential equations and their applications: Collège de France Seminar. Volume VI
H Brezis and J L Lions
110 Finite generalized quadrangles
S E Payne and J A Thas
111 Advances in nonlinear waves. Volume II
L Debnath
112 Topics in several complex variables
E Ramírez de Arellano and D Sundararaman
113 Differential equations, flow invariance and applications
N H Pavel
114 Geometrical combinatorics
F C Holroyd and R J Wilson
115 Generators of strongly continuous semigroups
J A van Casteren
116 Growth of algebras and Gelfand–Kirillov dimension
G R Krause and T H Lenagan
117 Theory of bases and cones
P K Kamthan and M Gupta
118 Linear groups and permutations
A R Camina and E A Whelan
119 General Wiener–Hopf factorization methods
F-O Speck
120 Free boundary problems: applications and theory, Volume III
A Bossavit, A Damlamian and M Fremond
121 Free boundary problems: applications and theory, Volume IV
A Bossavit, A Damlamian and M Fremond
122 Nonlinear partial differential equations and their applications: Collège de France Seminar. Volume VII
H Brezis and J L Lions
123 Geometric methods in operator algebras
H Araki and E G Effros
124 Infinite dimensional analysis–stochastic processes
S Albeverio
125 Ennio de Giorgi Colloquium
P Krée
126 Almost-periodic functions in abstract spaces
S Zaidman
127 Nonlinear variational problems
A Marino, L Modica, S Spagnolo and M Degiovanni
128 Second-order systems of partial differential equations in the plane
L K Hua, W Lin and C-Q Wu
129 Asymptotics of high-order ordinary differential equations
R B Paris and A D Wood
130 Stochastic differential equations
R Wu
131 Differential geometry
L A Cordero
132 Nonlinear differential equations
J K Hale and P Martinez-Amores
133 Approximation theory and applications
S P Singh
134 Near-rings and their links with groups
J D P Meldrum
135 Estimating eigenvalues with *a posteriori/a priori* inequalities
J R Kuttler and V G Sigillito
136 Regular semigroups as extensions
F J Pastijn and M Petrich
137 Representations of rank one Lie groups
D H Collingwood
138 Fractional calculus
G F Roach and A C McBride
139 Hamilton's principle in continuum mechanics
A Bedford
140 Numerical analysis
D F Griffiths and G A Watson
141 Semigroups, theory and applications. Volume I
H Brezis, M G Crandall and F Kappel
142 Distribution theorems of L-functions
D Joyner
143 Recent developments in structured continua
D De Kee and P Kaloni
144 Functional analysis and two-point differential operators
J Locker
145 Numerical methods for partial differential equations
S I Hariharan and T H Moulden
146 Completely bounded maps and dilations
V I Paulsen
147 Harmonic analysis on the Heisenberg nilpotent Lie group
W Schempp
148 Contributions to modern calculus of variations
L Cesari
149 Nonlinear parabolic equations: qualitative properties of solutions
L Boccardo and A Tesei
150 From local times to global geometry, control and physics
K D Elworthy

151 A stochastic maximum principle for optimal control of diffusions
U G Haussmann
152 Semigroups, theory and applications. Volume II
H Brezis, M G Crandall and F Kappel
153 A general theory of integration in function spaces
P Muldowney
154 Oakland Conference on partial differential equations and applied mathematics
L R Bragg and J W Dettman
155 Contributions to nonlinear partial differential equations. Volume II
J I Díaz and P L Lions
156 Semigroups of linear operators: an introduction
A C McBride
157 Ordinary and partial differential equations
B D Sleeman and R J Jarvis
158 Hyperbolic equations
F Colombini and M K V Murthy
159 Linear topologies on a ring: an overview
J S Golan
160 Dynamical systems and bifurcation theory
M I Camacho, M J Pacifico and F Takens
161 Branched coverings and algebraic functions
M Namba
162 Perturbation bounds for matrix eigenvalues
R Bhatia
163 Defect minimization in operator equations: theory and applications
R Reemtsen
164 Multidimensional Brownian excursions and potential theory
K Burdzy
165 Viscosity solutions and optimal control
R J Elliott
166 Nonlinear partial differential equations and their applications. Collège de France Seminar. Volume VIII
H Brezis and J L Lions
167 Theory and applications of inverse problems
H Haario
168 Energy stability and convection
G P Galdi and B Straughan
169 Additive groups of rings. Volume II
S Feigelstock
170 Numerical analysis 1987
D F Griffiths and G A Watson
171 Surveys of some recent results in operator theory. Volume I
J B Conway and B B Morrel
172 Amenable Banach algebras
J-P Pier
173 Pseudo-orbits of contact forms
A Bahri
174 Poisson algebras and Poisson manifolds
K H Bhaskara and K Viswanath
175 Maximum principles and eigenvalue problems in partial differential equations
P W Schaefer
176 Mathematical analysis of nonlinear, dynamic processes
K U Grusa
177 Cordes' two-parameter spectral representation theory
D F McGhee and R H Picard
178 Equivariant K-theory for proper actions
N C Phillips
179 Elliptic operators, topology and asymptotic methods
J Roe
180 Nonlinear evolution equations
J K Engelbrecht, V E Fridman and E N Pelinovski
181 Nonlinear partial differential equations and their applications. Collège de France Seminar. Volume IX
H Brezis and J L Lions
182 Critical points at infinity in some variational problems
A Bahri
183 Recent developments in hyperbolic equations
L Cattabriga, F Colombini, M K V Murthy and S Spagnolo
184 Optimization and identification of systems governed by evolution equations on Banach space
N U Ahmed
185 Free boundary problems: theory and applications. Volume I
K H Hoffmann and J Sprekels
186 Free boundary problems: theory and applications. Volume II
K H Hoffmann and J Sprekels
187 An introduction to intersection homology theory
F Kirwan
188 Derivatives, nuclei and dimensions on the frame of torsion theories
J S Golan and H Simmons
189 Theory of reproducing kernels and its applications
S Saitoh
190 Volterra integrodifferential equations in Banach spaces and applications
G Da Prato and M Iannelli
191 Nest algebras
K R Davidson
192 Surveys of some recent results in operator theory. Volume II
J B Conway and B B Morrel
193 Nonlinear variational problems. Volume II
A Marino and M K Murthy
194 Stochastic processes with multidimensional parameter
M E Dozzi
195 Prestressed bodies
D Iesan
196 Hilbert space approach to some classical transforms
R H Picard
197 Stochastic calculus in application
J R Norris
198 Radical theory
B J Gardner
199 The C* – algebras of a class of solvable Lie groups
X Wang

200 Stochastic analysis, path integration and dynamics
D Elworthy
201 Riemannian geometry and holonomy groups
S Salamon
202 Strong asymptotics for extremal errors and polynomials associated with Erdös type weights
D S Lubinsky
203 Optimal control of diffusion processes
V S Borkar
204 Rings, modules and radicals
B J Gardner
205 Two-parameter eigenvalue problems in ordinary differential equations
M Faierman
206 Distributions and analytic functions
R D Carmichael and D Mitrović
207 Semicontinuity, relaxation and integral representation in the calculus of variations
G Buttazzo
208 Recent advances in nonlinear elliptic and parabolic problems
P Bénilan, M Chipot, L Evans and M Pierre
209 Model completions, ring representations and the topology of the Pierce sheaf
A Carson
210 Retarded dynamical systems
G Stepan
211 Function spaces, differential operators and nonlinear analysis
L Paivarinta
212 Analytic function theory of one complex variable
C C Yang, Y Komatu and K Niino
213 Elements of stability of visco-elastic fluids
J Dunwoody
214 Jordan decompositions of generalised vector measures
K D Schmidt
215 A mathematical analysis of bending of plates with transverse shear deformation
C Constanda
216 Ordinary and partial differential equations Vol II
B D Sleeman and R J Jarvis
217 Hilbert modules over function algebras
R G Douglas and V I Paulsen
218 Graph colourings
R Wilson and R Nelson
219 Hardy-type inequalities
A Kufner and B Opic
220 Nonlinear partial differential equations and their applications. College de France Seminar Volume X
H Brezis and J L Lions
221 Workshop on dynamical systems
E Shiels and Z Coelho
222 Geometry and analysis in nonlinear dynamics
H W Broer anJ F Takens
223 Fluid dynamical aspects of combustion theory
M Onofri and A Tesei
224 Approximation of Hilbert space operators. Volume I. 2nd edition
D Herrero
225 Operator Theory: Proceedings of the 1988 GPOTS–Wabash conference
J B Conway and B B Morrel
226 Local cohomology and localization
J L Bueso Montero, B Torrecillas Jover and A Verschoren
227 Nonlinear waves and dissipative effects
D Fusco and A Jeffrey
228 Numerical analysis. Volume III
D F Griffiths and G A Watson
229 Recent developments in structured continua. Volume III
D De Kee and P Kaloni
230 Boolean methods in interpolation and approximation
F J Delvos and W Schempp
231 Further advances in twistor theory, Volume 1
L J Mason and L P Hughston
232 Further advances in twistor theory, Volume 2
L J Mason and L P Hughston
233 Geometry in the neighborhood of invariant manifolds of maps and flows and linearization
U Kirchgraber and K Palmer
234 Quantales and their applications
K I Rosenthal
235 Integral equations and inverse problems
V Petkov and R Lazarov
236 Pseudo-differential operators
S R Simanca
237 A functional analytic approach to statistical experiments
I M Bomze
238 Quantum mechanics, algebras and distributions
D Dubin and M Hennings
239 Hamilton flows and evolution semigroups
J Gzyl
240 Topics in controlled Markov chains
V S Borkar
241 Invariant manifold theory for hydrodynamic transition
S Sritharan
242 Lectures on the spectrum of $L^2(\Gamma\backslash G)$
F L Williams

M Faierman

University of the Witwatersrand, South Africa

Two-parameter eigenvalue problems in ordinary differential equations

Longman Scientific & Technical

Copublished in the United States with
John Wiley & Sons, Inc., New York

Longman Scientific & Technical,
Longman Group UK Limited,
Longman House, Burnt Mill, Harlow,
Essex CM20 2JE, England
and Associated Companies throughout the world.

Copublished in the United States with
John Wiley & Sons, Inc., 605 Third Avenue, New York, NY 10158

First published 1991

AMS Subject Classification: (Main) 34B25, 34B99, 47B50
(Subsidiary) 35J25, 35P10

ISSN 0269-3674

British Library Cataloguing in Publication Data
Faierman, M.
Two-parameter eigenvalue problems in ordinary
differential equations. – (Pitman research notes
in mathematics)
I. Title II. Series
515

ISBN 0-582-06117-2

Library of Congress Cataloging-in-Publication Data
Faierman, M.
Two-parameter eigenvalue problems in ordinary differential
equations / M. Faierman.
p. cm.— (Pitman research notes in mathematics series. ISSN 0269-3674; 205)
Includes bibliographical references.
1. Differential equations. 2.Spectral theory (Mathematics)
3. Differential operators. 4. Eigenvalues. I. Title. II. Series.
QA372.F325 1991
515′.352—dc20 91-18502
CIP

Printed and bound in Great Britain
by Biddles Ltd, Guildford and King's Lynn

Contents

Preface

CHAPTER 1 THE TWO-PARAMETER EIGENVALUE PROBLEM

1.1. Introduction 1
1.2. Preliminaries 1
1.3. Oscillation theory 2
1.4. Comments 8

CHAPTER 2 THE ASSOCIATED ELLIPTIC BOUNDARY VALUE PROBLEM

2.1. Introduction 10
2.2. Assumptions and definitions 10
2.3. The associated problem 11
2.4. Density and regularity results 15
2.5. The system (1.1-4) 30
2.6. Comments 34

CHAPTER 3 THE DECOMPOSITION OF $\mathcal{H}$

3.1. Introduction 36
3.2. Some facts from the theory of inner product spaces 36
3.3. Preliminaries 38
3.4. The subspace $\mathcal{H}_0$ 39
3.5. Comments 55

CHAPTER 4 THE RIGHT SEMI-DEFINITE CASE

4.1. Introduction 57
4.2. The problem (2.1) 57
4.3. The system (1.1-4) 60

4.4. Comments 65

CHAPTER 5 THE LEFT DEFINITE CASE

5.1. Introduction 68
5.2. The problem (2.1) 68
5.3. The system (1.1-4) 71
5.4. Comments 75

CHAPTER 6 THE NON-DEFINITE CASE

6.1. Introduction 77
6.2. Further facts from the theory of inner product spaces 77
6.3. The subspace V_0 78
6.4. The problem (2.1) 80
6.5. The system (1.1-4) 99
6.6. The principal subspaces of the problem (2.1) 114
6.7. The eigenfunction expansion 130
6.8. Comments 137

APPENDIX A A RIGHT DEFINITE TWO-PARAMETER EIGENVALUE PROBLEM INVOLVING COMPLEX POTENTIALS

A.1. Introduction 140
A.2. Preliminaries 140
A.3. The associated elliptic boundary value problem 142
A.4. The Hilbert space $\mathcal{H}_\omega$ 144
A.5. The system (1.1-4) 147
A.6. Comments 151

REFERENCES 153

Preface

Multiparameter spectral theory began with the oscillation theorem of Klein [**121**], wherein Lamé's differential equation, which involves two spectral parameters, was considered over two disjoint intervals and the parameters were to be chosen so that the differential equation admits in each of these intervals a non-trivial solution vanishing at the endpoints. Extensions of Klein's original results to more general multiparameter Sturm–Liouville problems were given by Bôcher [**44 - 46**] and Klein [**122**] towards the end of the nineteenth century and by Dixon [**61**], Hilb [**105**], Hilbert [**106**], Klein [**122**], and various students of Hilbert in the first two decades of the twentieth century. Bearing in mind the great variety of eigenvalue problems involving several parameters, there was a break in the 1920's from the traditional problems studied by previous investigators in that there appeared during this period the works of Camp [**50,51**], Carmichael [**52 - 54**], Doole [**62**], and Pell [**137**] on multiparameter eigenvalue problems no longer restricted to ordinary differential equations of the second order. Apart from the important works of Arscott [**7**] and Cordes [**56 - 58**], multiparameter spectral theory then remained relatively neglected from the 1930's until almost the mid 1960's at which time there began to appear the fundamental works of Atkinson [**9 - 11**] proposing a theory which would completely subsume all previous investigations.

In the theory of Atkinson one has k operators involving k spectral parameters, with each operator acting in a distinct Hilbert space, and with the operators being linked only through the spectral parameters. These operators then induce in the tensor product of the Hilbert spaces concerned k commutating operators involving k parameters; and it is by a study of the joint spectra, eigenvectors, etc., of these latter operators that basic results concerning the original multiparameter problem are established. This systematic theory of Atkinson rekindled a renewed interest in eigenvalue problems involving several parameters and has indeed laid the foundations of most of the advances which have taken place in this area of investigation over the last two decades.

However, it is felt by this author that, because of its generality, Atkinson's theory does not necessarily produce in all cases the most complete results. In particular, for the case of problems involving ordinary differential operators of the second order, we feel that the method proposed by Hilbert, i.e., the reduction to an elliptic boundary value problem, is more effective. Accordingly, in this book we shall justify this sentiment by considering the following two problems associated with such operators, namely: (1) a two-parameter eigenvalue problem which is not right definite, to which the main part of the book is devoted, and (2) a right definite two-parameter eigenvalue problem involving complex potentials, which is treated in Appendix A. Then using techniques from the theory of elliptic boundary value problems, we shall derive results not hitherto established by other methods. Moreover, although we have restricted ourselves here to ordinary differential operators of the second order involving two parameters, a further consequence of our results, and this should be of particular interest to specialists in multiparameter spectral theory, is that they indicate what is to be expected in more general situations. Since our methods are based upon the theory of elliptic boundary value problems, the book should also be of interest to workers in this area who are especially concerned with problems involving an indefinite weight function since §§5.2 and 6.4 contain results pertaining to such problems which complement the known ones.

The author wishes to express his gratitude to Professor G.F. Roach of the University of Strathclyde for his invitation to contribute a book to the Pitman Research Notes in Mathematics series and would like to point out that many of the results contained in this book are a consequence of our mutual collaboration. We also wish to thank Professor P.A. Binding of the University of Calgary for making available preprints of his latest works.

Finally, of course, the author gratefully acknowledges the expertise of Mrs. Ingrid Eitzen who prepared the complete typescript.

CHAPTER 1

The two-parameter eigenvalue problem

1.1. Introduction

We have already stated in the preface that the main part of this book will be devoted to the establishment of some basic facts concerning the eigenvalues and eigenfunctions of a two-parameter eigenvalue problem which is not right definite. Accordingly in §1.2 below we introduce the two-parameter eigenvalue problem in question. Moreover, since multiparameter spectral theory originated with the oscillation theorem of Klein [**121**], it will be fitting to begin our work with some oscillation theory, and hence in §1.3 we establish some results of this nature which we require in the sequel. Finally, in §1.4 we conclude the chapter with some comments.

1.2. Preliminaries

We will be concerned in this book with the simultaneous two-parameter systems

$$(p_1(x_1)y_1')' + \Big(\lambda_1 A_1(x_1) - \lambda_2 B_1(x_1) - q_1(x_1)\Big)y_1 = 0, \tag{1.1}$$

$$0 \le x_1 \le 1, \quad ' = d/dx_1,$$

$$y_1(0)\cos\alpha_1 - p_1(0)y_1'(0)\sin\alpha_1 = 0, \quad 0 \le \alpha_1 < \pi, \tag{1.2a}$$

$$y_1(1)\cos\beta_1 - p_1(1)y_1'(1)\sin\beta_1 = 0, \quad 0 < \beta_1 \le \pi, \tag{1.2b}$$

and

$$(p_2(x_2)y_2')' + \Big(-\lambda_1 A_2(x_2) + \lambda_2 B_2(x_2) - q_2(x_2)\Big)y_2 = 0, \tag{1.3}$$

$$0 \le x_2 \le 1, \quad ' = d/dx_2,$$

$$y_2(0)\cos\alpha_2 - p_2(0)y_2'(0)\sin\alpha_2 = 0, \quad 0 \le \alpha_2 < \pi, \tag{1.4a}$$

$$y_2(1)\cos\beta_2 - p_2(1)y_2'(1)\sin\beta_2 = 0, \quad 0 < \beta_2 \le \pi, \tag{1.4b}$$

where throughout this chapter we suppose that in $0 \leq x_r \leq 1$ $(r = 1, 2)$, p_r is positive and Lipschitz continuous, A_r and B_r are real–valued and continuous, while q_r is real–valued and essentially bounded.

Writing λ for (λ_1, λ_2), we call $\lambda^\star$ an eigenvalue of the system (1.1–4) if for $\lambda = \lambda^\star$, $(1.2r - 1)$ has a non-trivial solution, say $y_r(x_r, \lambda^\star)$, satisfying $(1.2r)$ for $r = 1, 2$. The product $\prod_{r=1}^{2} y_r(x_r, \lambda^\star)$ is called an eigenfunction of (1.1-4) corresponding to $\lambda^\star$. Then for the remainder of this book, with the exception of Appendix A, it will be our objective to establish some basic facts concerning the eigenvalues and eigenfunctions of the system (1.1-4) under the following suppositions.

Assumption 1.1. It will henceforth be supposed that:

(i) $\omega = A_1 B_2 - A_2 B_1$ vanishes in I^2, but is not identically zero, where

$$I^2 = \left\{ (x_1, x_2) \in \mathbb{R}^2 | 0 \leq x_r \leq 1 \text{ for } r = 1, 2 \right\},$$

(ii) there exist real numbers $\{\tau_i\}_1^2$ such that

$$\tau_1 A_r(x_r) + \tau_2 B_r(x_r) > 0 \text{ in } 0 \leq x_r \leq 1 \text{ for } r = 1, 2.$$

Part (ii) of Assumption 1.1 is known as the ellipticity condition; and it is precisely this condition which will enable us to establish our main results by an analysis of the spectral properties of an associated elliptic boundary value problem. We will discuss this matter in greater detail in Chapter 2; however for the moment we will content ourselves with establishing some preliminary results concerning the eigenvalues and eigenfunctions of the system (1.1-4).

1.3. Oscillation theory

Guided by future requirements we are now going to establish some facts concerning the eigenvalues and eigenfunctions of the system (1.1-4). Accordingly, if $\lambda^\star = (\lambda_1^\star, \lambda_2^\star)$ is an eigenvalue of the system (1.1-4), then we say that $\lambda^\star$ is real if both $\lambda_1^\star$ and $\lambda_2^\star$ are real. Let us also denote by $\phi_r(x_r, \lambda)$ the solution of $(1.2r - 1)$ satisfying

$$\phi_r(0, \lambda) = \sin \alpha_r, \; p_r(0)\phi_r'(0, \lambda) = \cos \alpha_r \text{ for } r = 1, 2.$$

Theorem 1.1. The real eigenvalue of the system (1.1-4) form a denumerably infinite subset of $\mathbb{R}^2$ having no finite points of accumulation. In particular, there exists the integer $n_1 \geq 0$ such that if (n,m) is any tuple of non-negative integers for which $n \geq n_1$, then to this tuple there corresponds a finite number of real eigenvalues of the system (1.1-4) such that at each such eigenvalue the solution of (1.1) satisfying (1.2) has exactly n zeros in $0 < x_1 < 1$ and the solution of (1.3) satisfying (1.4) has exactly m zeros in $0 < x_2 < 1$. If $n_1 > 0$ and $0 \leq n < n_1$, then there exists the integer $m(n)$, where $m(0) \geq m(1) \geq \cdots \geq m(n_1 - 1) \geq 1$, such that for any integer $m \geq m(n)$, there corresponds to the tuple (n,m) a finite number of real eigenvalues of (1.1-4) with the above oscillation result holding at each such eigenvalue. However, if $0 \leq m < m(n)$, then there exist no real eigenvalues corresponding to the tuple (n,m), that is, at no real eigenvalue of the system (1.1-4) does the solution of (1.1) satisfying (1.2) have exactly n zeros in $0 < x_1 < 1$ and the solution of (1.3) satisfying (1.4) have exactly m zeros in $0 < x_2 < 1$.

Proof. If we let $\tau_j^\dagger = \tau_j(\tau_1^2 + \tau_2^2)^{-1/2}$ for $j = 1,2$ and introduce the transformation

$$\lambda_1 = \tau_1^\dagger \mu + \tau_2^\dagger \nu, \quad \lambda_2 = -\tau_2^\dagger \mu + \tau_1^\dagger \nu,$$

then (1.1) and (1.3) become

$$(p_1(x_1)y_1')' + \left(\mu A_1^\dagger(x_1) - \nu B_1^\dagger(x_1) - q_1(x_1)\right) y_1 = 0,\ 0 \leq x_1 \leq 1, \tag{1.5}$$

and

$$(p_2(x_2)y_2')' + \left(-\mu A_2^\dagger(x_2) + \nu B_2^\dagger(x_2) - q_2(x_2)\right) y_2 = 0,\ 0 \leq x_2 \leq 1, \tag{1.6}$$

respectively, where $A_r^\dagger(x_r) > 0$ in $0 \leq x_r \leq 1$ for $r = 1,2$. In order to prove the theorem it will be necessary to establish some information concerning each of the systems (1.5), (1.2) and (1.6), (1.4); and to this end we need the following definitions. Let γ_1 and γ_2 denote the supremum and infimum, respectively, of $B_1^\dagger(x_1)/A_1^\dagger(x_1)$ in $0 \leq x_1 \leq 1$ and let $\gamma_1^\star$ and $\gamma_2^\star$ denote the supremum and infimum, respectively, of $B_2^\dagger(x_2)/A_2^\dagger(x_2)$ in $0 \leq x_2 \leq 1$. Let $\psi_r = \tan^{-1}\gamma_r$, $\psi_r^\star = \tan^{-1}\gamma_r^\star$ for $r = 1,2$, $\psi_3 = \psi_1 + \pi$, and $\psi_3^\star = \psi_2^\star + \pi$, where the principal branch of the inverse tangent is taken.

Let us firstly fix our attention upon the system (1.5), (1.2). Then we know from the Sturm theory that for each real ν, the totality of values of μ for which (1.5) has a non-trivial solution satisfying (1.2) forms a countably infinite subset of real numbers which we denote by $\mu_n(\nu), n \geq 0$, where $\mu_0(\nu) < \mu_1(\nu) < \cdots, \mu_n(\nu) \to \infty$ as $n \to \infty$ and the solution corresponding to $\mu_n(\nu)$ has precisely n zeros in $0 < x_1 < 1$. Thus we see that the $\mu_n(\nu)$ determine a countably infinite number of disjoint curves in the (ν,μ)-plane which we denote by $s_n, n \geq 0$. The eigenvalue curves s_n have been the subject of much investigation and we now collect the following well known facts concerning them (see §1.4 below). For each $n, \mu_n(\nu)$ is analytic in $-\infty < \nu < \infty$ and at each point of this interval

$$d\mu_n(\nu)/d\nu = \left(\int_0^1 B_1^\dagger(x_1)\chi_1^2(x_1,\nu,\mu_n(\nu))dx_1\right) \times$$
$$\times \left(\int_0^1 A_1^\dagger(x_1)\chi_1^2(x_1,\nu,\mu_n(\nu))dx_1\right)^{-1},$$

where $\chi_1(x_1,\nu,\mu) = \phi_1(x_1,\lambda(\nu,\mu))$. If ψ denotes the angle which a ray emanating from the origin in the (ν,μ)-plane makes with the positive ν-axis, if ϵ is a number satisfying $0 < \epsilon < \min\{-\psi_1 + \pi/2, \psi_2 + \pi/2\}$, and n is any non-negative integer, then there exists the positive number $\nu_n(\epsilon)$ such that $(\nu, \mu_n(\nu))$ lies in the sector $\psi_2 - \epsilon < \psi < \psi_2 + \epsilon$ if $\nu \geq \nu_n(\epsilon)$ and in the sector $\psi_3 - \epsilon < \psi < \psi_3 + \epsilon$ if $\nu \leq -\nu_n(\epsilon)$. Lastly, if d_n denotes the distance of s_n to the origin of the (ν,μ)-plane, then $d_n \to \infty$ as $n \to \infty$.

In Figure 1 below we illustrate the above discussion by means of a sketch. Here it is supposed that $B_1^\dagger(x_1)$ assumes both positive and negative values in $0 \leq x_1 \leq 1$.

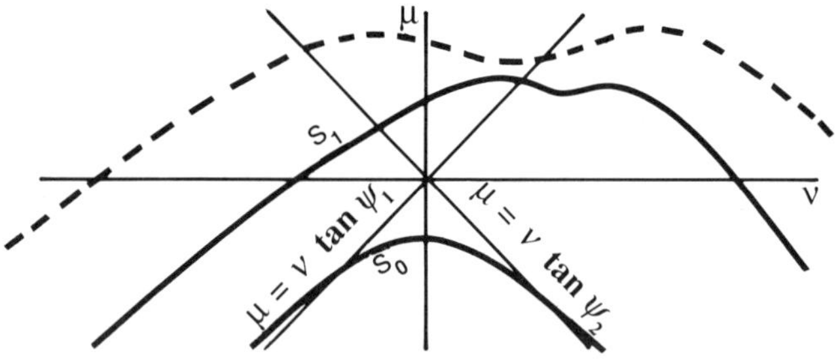

Figure 1.

Turning next to the system (1.6), (1.4), we know that for each real ν the totality of values of μ for which (1.6) has a non-trivial solution satisfying (1.4) forms a countably infinite subset of real numbers which we denote by $\mu_n^\star(\nu)$, $n \geq 0$, where $\mu_0^\star(\nu) > \mu_1^\star(\nu) > \cdots, \mu_n^\star(\nu) \to -\infty$ as $n \to \infty$, and the solution corresponding to $\mu_n^\star(\nu)$ has precisely n zeros in $0 < x_2 < 1$. Moreover, for each n, $\mu_n^\star(\nu)$ is analytic in $-\infty < \nu < \infty$ and at each point of this interval

$$d\mu_n^\star(\nu)/d\nu = \left(\int_0^1 B_2^\dagger(x_2)\chi_2^2(x_2,\nu,\mu_n^\star(\nu))dx_2\right) \times$$
$$\times \left(\int_0^1 A_2^\dagger(x_2)\chi_2^2(x_2,\nu,\mu_n^\star(\nu))dx_2\right)^{-1},$$

where $\chi_2(x_2,\nu,\mu) = \phi_2(x_2,\lambda(\nu,\mu))$. Thus the $\mu_n^\star(\nu)$ determine a countably infinite number of disjoint analytic curves in the (ν,μ)-plane which we denote $s_n^\star$, $n \geq 0$. Fixing our attention upon this plane and with ψ denoting angle as introduced above, we know that for any ϵ satisfying $0 < \epsilon < \min\{-\psi_1^\star + \pi/2, \psi_2^\star + \pi/2\}$ and any non-negative integer n there exists the positive number $\nu_n^\star(\epsilon)$ such that $(\nu,\mu_n^\star(\nu))$ lies in the sector $\psi_1^\star - \epsilon < \psi < \psi_1^\star + \epsilon$ if $\nu \geq \nu_n^\star(\epsilon)$ and in the sector $\psi_3^\star - \epsilon < \psi < \psi_3^\star + \epsilon$ if $\nu \leq -\nu_n^\star(\epsilon)$. Lastly, if $d_n^\star$ denotes the distance of $s_n^\star$ to the origin in the (ν,μ)-plane, then $d_n^\star \to \infty$ as $n \to \infty$.

In Figure 2 below we illustrate the above discussion by means of a sketch. Here it is supposed that $B_2^\dagger(x_2)$ is positive and $B_2^\dagger(x_2)/A_2^\dagger(x_2)$ is not constant in $0 \leq x_2 \leq 1$.

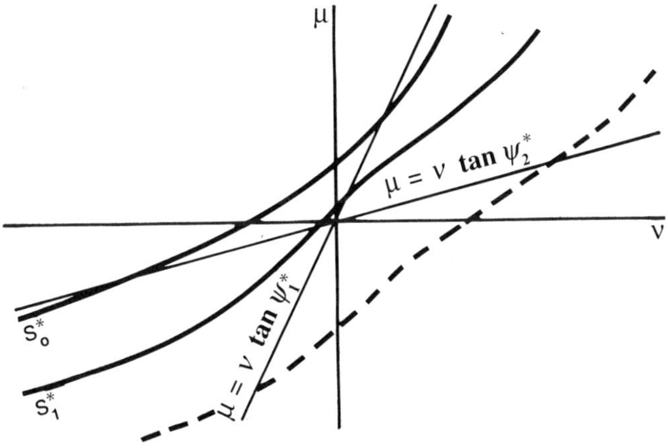

Figure 2.

It is clear that the real eigenvalues of the system (1.1-4) are precisely the images of the points of intersection of the s_n and the $s_m^\star$ under an orthogonal transformation. Thus in light of the foregoing results we see that in order to prove the theorem we have to show that the s_n and $s_m^\star$ do indeed have points of intersection, pick out those s_n and $s_m^\star$ which do intersect and show that an s_n can only intersect an $s_m^\star$ in at most a finite number of points. As a consequence of this last statement and our above results, it will then follow that any compact subset of $\mathbb{R}^2$ can only contain at most a finite number of real eigenvalues of the system (1.1-4). Accordingly, if we observe that $A_1^\dagger(x_1)B_2^\dagger(x_2) - A_2^\dagger(x_2)B_1^\dagger(x_1) = \omega(x)$ in I^2, where $x = (x_1, x_2)$, then we conclude from Assumption 1.1 that either

(a) $B_2^\dagger(x_2)/A_2^\dagger(x_2)$ is not constant in $0 \le x_2 \le 1$ and precisely one of the following conditions hold:

(i) $\psi_1 = \psi_2^\star < \psi_1^\star$, (ii) $\psi_2^\star < \psi_1 < \psi_1^\star$, (iii) $\psi_2 < \psi_1^\star \le \psi_1$,

(iv) $\psi_2 = \psi_1^\star \le \psi_1$,

or

(b) $B_2^\dagger(x_2)/A_2^\dagger(x_2)$ is constant in $0 \le x_2 \le 1$ and precisely one of the following conditions hold:

(i) $\psi_2 < \psi_1 = \psi_1^\star$, (ii) $\psi_2 < \psi_1^\star < \psi_1$, (iii) $\psi_2 = \psi_1^\star < \psi_1$.

With n, m denoting any pair of non-negative integers, let us firstly fix our attention upon case (a). Then it is clear from the above results that

1. $d\mu_m^\star(\nu)/d\nu - d\mu_n(\nu)/d\nu > 0$ in $-\infty < \nu < \infty$

   and $\mu_m^\star(\nu) - \mu_n(\nu) \to \infty$ as $\nu \to \infty$ if condition (i) holds,

2. $\mu_m^\star(\nu) - \mu_n(\nu) \to \infty$ as $\nu \to \pm\infty$ if conditions (ii) or (iii) hold,

3. $d\mu_m^\star(\nu)/d\nu - d\mu_n(\nu)/d\nu < 0$ in $-\infty < \nu < \infty$

   and $\mu_n^\star(\nu) - \mu_n(\nu) \to \infty$ as $\nu \to -\infty$ if condition (iv) holds.

Turning next to case (b), it is clear that

1. $d\mu_m^\star(\nu)/d\nu - d\mu_n(\nu)/d\nu > 0$ in $-\infty < \nu < \infty$

   and $\mu_m^\star(\nu) - \mu_n(\nu) \to \infty$ as $\nu \to \infty$ if condition (i) holds,

2. $\mu_m^\star(\nu) - \mu_n(\nu) \to \infty$ as $\nu \to \pm\infty$ if condition (ii) holds,

3. $d\mu^{\star}_m(\nu)/d\nu - d\mu_n(\nu)/d\nu < 0$ in $-\infty < \nu < \infty$

and $\mu^{\star}_m(\nu) - \mu_n(\nu) \to \infty$ as $\nu \to -\infty$ if condition (iii) holds.

Since $\mu_n(0) \geq \mu^{\star}_0(0)$ if n is sufficiently large and since $\mu^{\star}_0(\nu) - \mu_n(\nu) \to \infty$ as $\nu \to \infty$ or as $\nu \to -\infty$, we conclude that s_n intersects $s^{\star}_0$ for all n sufficiently large. Hence there exists the integer $n_1 \geq 0$ such that s_{n_1} intersects $s^{\star}_0$, while if $n_1 > 0$ and $0 \leq n < n_1$, then s_n does not intersect $s^{\star}_0$. If n, m is any pair of non-negative integers such that $n \geq n_1$, then it follows that $\mu_n(\nu^{\dagger}) \geq \mu^{\star}_m(\nu^{\dagger})$ for some $\nu^{\dagger}$ in $(-\infty, \infty)$ and since $\mu^{\star}_m(\nu) - \mu_n(\nu) \to \infty$ as $\nu \to \infty$ or as $\nu \to -\infty$, we therefore conclude that s_n and $s^{\star}_m$ intersect. Moreover, if we refer to the discussion of the previous paragraph and suppose that either we are dealing with case (a) and precisely one of the conditions (i) or (iv) hold or we are dealing with case (b) and precisely one of the conditions (i) or (iii) hold, then it follows from the fact that $d\mu^{\star}_m(\nu)/d\nu - d\mu_n(\nu)/d\nu \neq 0$ in $-\infty < \nu < \infty$, that s_n intersects $s^{\star}_m$ in precisely one point. If, however, we are dealing with any one of the remaining cases considered in the previous paragraph, then it follows from the facts that $\mu_n(\nu)$ and $\mu^{\star}_m(\nu)$ are analytic in $-\infty < \nu < \infty$ and $\mu^{\star}_m(\nu) - \mu_n(\nu) \to \infty$ as $\nu \to \pm\infty$, that s_n intersects $s^{\star}_m$ in a finite number of points. Next, let us suppose that $n_1 > 0$ and $0 \leq n < n_1$. Then since $\mu^{\star}_m(0) \leq \mu_n(0)$ if m is sufficiently large, it follows from the results of the previous paragraph that $s^{\star}_m$ intersects s_n for all m sufficiently large. Thus there exists the integer $m(n) \geq 1$ such that $s^{\star}_{m(n)}$ intersects s_n, while if $0 \leq m < m(n)$, then $s^{\star}_m$ does not intersect s_n. If $m \geq m(n)$, then it is clear that $\mu_n(\nu^{\dagger}) \geq \mu^{\star}_m(\nu^{\dagger})$ for some $\nu^{\dagger}$ in $(-\infty, \infty)$, and since $\mu^{\star}_m(\nu) - \mu_n(\nu) \to \infty$ as $\nu \to \infty$ or as $\nu \to -\infty$, we conclude that $s^{\star}_m$ and s_n intersect. Furthermore, all the results derived above concerning the number of points of intersection of $s^{\star}_m$ and s_n when $n \geq n_1$ carry over without change to the case under consideration here. Finally, since it is clear that $m(0) \geq m(1) \geq \cdots \geq m(n_1 - 1) \geq 1$, the proof of the theorem is complete.

<u>Remark 1.1.</u> We have shown in the above proof that when $\omega \geq 0$ or $\omega \leq 0$ in I^2, then there corresponds precisely one real eigenvalue of the system (1.1-4) to each tuple of non-negative integers satisfying the conditions of the theorem.

If we henceforth write x for (x_1, x_2) and let $\psi^{\star}(x, \lambda) = \prod_{r=1}^{2} \phi_r(x_r, \lambda)$, then

Theorem 1.2. Let $\lambda^\dagger = (\lambda_1^\dagger, \lambda_2^\dagger)$ and $\lambda^\# = (\lambda_1^\#, \lambda_2^\#)$ be any two distinct eigenvalues of the system (1.1-4) and put $\Psi^\dagger(x) = \psi^\star(x, \lambda^\dagger), \Psi^\#(x) = \psi^\star(x, \lambda^\#)$. Then $\int_{I^2} \omega \Psi^\dagger \Psi^\# dx = 0$.

Proof. Let us denote the equation (1.1) by (1.1†) when $\lambda_j = \lambda_j^\dagger$ for $j = 1,2$ and $y_1 = \phi_1(x_1, \lambda^\dagger)$, and by (1.1#) when $\lambda_j = \lambda_j^\#$ for $j = 1,2$ and $y_1 = \phi_1(x_1, \lambda^\#)$. Then if we multiply (1.1†) by $\phi_1(x_1, \lambda^\#)$, subtract from it (1.1#) multiplied by $\phi_1(x_1, \lambda^\dagger)$, and integrate over [0,1], we obtain

$$(\lambda_1^\dagger - \lambda_1^\#) \int_0^1 A_1(x_1)\phi_1(x_1, \lambda^\dagger)\phi_1(x_1, \lambda^\#) dx_1 -$$

$$-(\lambda_2^\dagger - \lambda_2^\#) \int_0^1 B_1(x_1)\phi_1(x_1, \lambda^\dagger)\phi_1(x_1, \lambda^\#) dx_1 = 0.$$

Similarly we can show that

$$(\lambda_1^\dagger - \lambda_1^\#) \int_0^1 A_2(x_2)\phi_2(x_2, \lambda^\dagger)\phi_2(x_2, \lambda^\#) dx_2 -$$

$$-(\lambda_2^\dagger - \lambda_2^\#) \int_0^1 B_2(x_2)\phi_2(x_2, \lambda^\dagger)\phi_2(x_2, \lambda^\#) dx_2 = 0.$$

Hence since $\lambda^\dagger \neq \lambda^\#$, we arrive at the assertion of the theorem.

1.4. Comments

We have stated in §1.2 that in the main part of this book we will be concerned with the eigenvalue problem (1.1-4) under Assumption 1.1. This problem was suggested to the author by F.V. Atkinson in a private communication. Historically Assumption 1.1 was preceded by the requirement that $\omega \neq 0$ in I^2 (cf. Klein [**121**]), and a somewhat restricted version of the assumption appears to have been firstly considered by Hilbert [**106**, Chapter 21]. For further investigations of general multiparameter eigenvalue problems involving unbounded operators under conditions which are in some way related to Assumption 1.1, we refer to Binding [**17-21, 24**], Binding and Browne [**28**], Binding, Källström, and Sleeman [**37**], Binding and Seddighi [**38**], Binding and Volkmer [**40**], Faierman [**74-83**], Faierman and Roach [**89,90,92**], Källström and Sleeman [**116,117**], Sleeman [**144-147**], and Volkmer [**152**].

The proof of Theorem 1.1 follows closely the arguments of [**74**] wherein use is made of the results of Richardson [**140,141**] and the author [**71**] concerning the properties of the eigenvalue curves s_n and $s^\star_m$. For related oscillation results and generalizations we refer to the references cited in the previous paragraph, while for further results concerning the geometry of the eigenvalue curves for general two-parameter eigenvalue problems and for extensions to k-parameter problems we refer to Allegretto and Mingarelli [**6**], Binding [**23**], Binding and Browne [**26,27,29-32**], Binding, Browne, and Turyn [**33-36**], Faierman [**70**], and Turyn [**150**].

One striking consequence of Theorem 1.1 is the possible deficiency of real eigenvalues, i.e., there may exist no real eigenvalues of the system (1.1-4) corresponding to a certain tuple of non-negative integers (in the sense of the theorem). Furthermore, if there is a deficiency of real eigenvalues, then there arises the question of whether this loss is compensated for by the appearance of non-real eigenvalues. Another consequence of the theorem is the problem of ascertaining the number of distinct real eigenvalues of (1.1-4) corresponding to a given tuple of non-negative integers. Although not much work has been devoted to the study of such topics as far as multiparameter spectral theory is concerned, they have certainly been the subject of much investigation for the 1-parameter analogue of the system (1.1-4) (i.e., the weighted Sturm-Liouville problem) and other 1-parameter systems. For recent relevant investigations we refer to Allegretto and Mingarelli [**6**], Atkinson and Jabon [**12**], Binding and Browne [**30**], Everitt, Kwong, and Zettl [**66**], and Mingarelli [**133**].

CHAPTER 2

The associated elliptic boundary value problem

2.1. Introduction

The ellipticity condition of Assumption 1.1 suggests that one way of dealing with the eigenvalue problem (1.1-4) is to proceed in the opposite direction to that of separation of variables, that is, to go back to the original elliptic boundary value problem from which (1.1-4) arose, to establish some basic facts concerning the eigenvalues and eigenvectors of the elliptic problem, and then to arrive at our main results by demonstrating their relationship to the eigenvalues and eigenfunctions of (1.1-4). This is the method that was initiated by Hilbert [**106**] and which will form the basis of our work. Accordingly, in §2.2 of this chapter we introduce assumptions and definitions which are required in the sequel, while in §2.3 we define the elliptic boundary value problem associated with the eigenvalue problem (1.1-4) and various related results are established. In §2.4 we discuss the regularity of solutions of this problem, while in §2.5 we establish our first basic results connecting the elliptic problem with the eigenvalue problem (1.1-4). Finally, in §2.6 we conclude the chapter with some comments.

2.2. Assumptions and definitions

In order to ensure that the eigenfunctions of the elliptic boundary value problem have sufficient regularity to enable our method to work, we require

Assumption 2.1. In addition to the hypotheses concerning the p_r, A_r, B_r, and q_r given in §1.2 of Chapter 1, we also suppose from now on that A_r and B_r are Lipschitz continuous in $0 \leq x_r \leq 1$ for $r = 1, 2$.

In order to introduce our elliptic boundary value problem, we also require

Assumption 2.2. We henceforth suppose that in ($1.2r$-1), $A_r(x_r) > 0$ in $0 \leq x_r \leq 1$ for $r = 1, 2$.

We would mention that Assumption 2.2 in no way diminishes the generality of our work since we know from the proof of Theorem 1.1 that the conditions guaranteed by this assumption can always be achieved if necessary by means of a change of parameters of the form $\lambda_r = \sum_{s=1}^{2} u_{rs}\lambda'_s$, $r = 1, 2$, where $U = (u_{rs})$ is an orthogonal matrix.

In the sequel we let Ω and Γ denote the interior and boundary, respectively, of I^2 and for $r = 1, 2$ (resp. $r = 3, 4$), let Γ_r denote the face of Ω : $x_r = 0$, $0 < x_{3-r} < 1$ (resp. $x_{r-2} = 1$, $0 < x_{5-r} < 1$). On Γ_r we define functions $b_r(x)$ and $\sigma_r(x)$ as follows (see (1.2) and (1.4)):

(1) if $1 \le r \le 2$, then we let

$$b_r(x) = A_s(x_s)p_r(0), \sigma_r(x) = A_s(x_s)\cot\alpha_r \text{ on } \Gamma_r \text{ if } \alpha_r \neq 0, \text{ where } s = 3 - r,$$

$$b_r(x) = 0, \quad \sigma_r(x) = 1 \text{ on } \Gamma_r \text{ if } \alpha_r = 0;$$

(2) if $3 \le r \le 4$, then we let

$$b_r(x) = A_s(x_s)p_{r-2}(1), \; \sigma_r(x) = -A_s(x_s)\cot\beta_{r-2} \text{ on } \Gamma_r \text{ if } \beta_{r-2} \neq \pi,$$

where $s = 5 - r$,

$$b_r(x) = 0, \quad \sigma_r(x) = 1 \text{ on } \Gamma_r \text{ if } \beta_{r-2} = \pi.$$

Referring again to (1.2) and (1.4), suppose that for at least one j we have $\alpha_j = 0$ or $\beta_j = \pi$ and for at least one j, $\alpha_j \neq 0$ or $\beta_j \neq \pi$. Then for this case we let Γ'' denote the union of those Γ_r for which $\alpha_r = 0$ if $1 \le r \le 2$ and $\beta_{r-2} = \pi$ if $3 \le r \le 4$, and let $\Gamma' = \Gamma \backslash \Gamma''$.

We henceforth say that a function $f \in C^1(I^2)$ satisfies the boundary conditions (1.2) and (1.4) if it satisfies (1.2a) on $x_1 = 0$, (1.2b) on $x_1 = 1$, (1.4a) on $x_2 = 0$, and (1.4b) on $x_2 = 1$. We also let $\mathcal{H} = L^2(\Omega)$ and denote by $(\ ,\ )$ and $\|\ \ \|$ the inner product and norm, respectively, in $\mathcal{H}$. For $q \ge 1$, k a non-negative integer, and G any open subset of $\mathbb{R}^2$, we let $W^{k,q}(G)$ denote the usual Sobolev space of order k related to $L^q(G)$, put $H^k(G) = W^{k,2}(G)$, and let $(\ ,\ )_{k,G}$ and $\|\ \ \|_{k,G}$ denote the inner product and norm, respectively, in $H^k(G)$. Lastly, if A is a linear operator mapping one vector space into another, then we let $D(A)$ = domain of A, $R(A)$ = range of A, and $N(A)$ = ker A.

2.3. The associated problem

Let $a_1(x) = p_1(x_1)A_2(x_2)$, $a_2(x) = p_2(x_2)A_1(x_1)$, and $q(x) = q_1(x_1)A_2(x_2) + q_2(x_2)A_1(x_1)$

for $x \in I^2$. Then if we proceed in a formal manner and multiply (1.1) by $A_2(x_2)y_2$, (1.3) by $A_1(x_1)y_1$, and add, we arrive at the boundary value problem

$$Lu = \mu\omega(x)u \text{ in } \Omega, \tag{2.1a}$$

$$b_r(x)\partial u/\partial\nu_r + \sigma_r(x)u = 0 \text{ on } \Gamma_r \text{ for } r = 1,\ldots,4, \tag{2.1b}$$

where L denotes the elliptic operator

$$-\sum_{r=1}^{2} D_r a_r(x) D_r + q(x),$$

$D_r = \partial/\partial x_r$ for $r = 1,2$, $u = y_1 y_2$, $\mu = \lambda_2$, and ν_r denotes the unit exterior normal to Γ_r. It is precisely (2.1) which, at least formally, is the elliptic boundary value problem associated with the system (1.1-4); and since it is our intention to deal with this problem from a variational point of view, we require some preliminary results.

Accordingly, it will be convenient for us to refer to (2.1) as: (i) BVP I if $\alpha_r = 0$ and $\beta_r = \pi$ for $r = 1,2$, (ii) BVP II if $\alpha_r \neq 0$ and $\beta_r \neq \pi$ for $r = 1,2$, and (iii) BVP III otherwise.

Definition 2.1. If we are dealing with BVP I (resp. BVP III), then let V denote the subspace of $H^1(\Omega)$ consisting of those elements which vanish, in the sense of trace, on all the Γ_r (resp. on those Γ_r for which $\Gamma_r \subset \Gamma''$). If we are dealing with BVP II, then we let $V = H^1(\Omega)$.

In §2.4 we will give another characterization of V; however for the moment let us note that V is a closed subspace of $H^1(\Omega)$. Next let

$$B^\dagger(u,v) = \sum_{r=1}^{2}(D_r u, a_r D_r v) + (u, qv) \text{ for } u,v \in H^1(\Omega).$$

Then in V we introduce the sesquilinear form $B(u,v)$ in the following way.

Definition 2.2. If we are dealing with BVP I, then let $B(u,v) = B^\dagger(u,v)$ for $u,v \in V$. If we are dealing with BVP II, then let

$$B(u,v) = B^\dagger(u,v) + \int_\Gamma \sigma(tru)(\overline{trv})ds \text{ for } u,v \in V,$$

where Γ is assumed to be oriented in the positive sense, s denotes arc length, trf denotes the trace of f on Γ, and $\sigma(x) = \sigma_r(x)$ if $x \in \Gamma_r$. If we are dealing with BVP III, then let

$$B(u,v) = B^\dagger(u,v) + \int_{\Gamma'} \sigma(tru)(\overline{trv})ds \quad \text{for} \quad u,v \in V,$$

where $\sigma(x) = \sigma_r(x)$ if $x \in \Gamma_r$ and $\Gamma_r \subset \Gamma'$.

It is clear that the form B is symmetric, that is, $B(v,u) = \overline{B(v,u)}$ for $u,v \in V$. We are now going to show that B is coercive over V, that is, there exist constants $c_0 > 0$ and $c_1 \geq 0$ such that

$$B(u,u) \geq c_0\|u\|^2_{1,\Omega} - c_1\|u\|^2 \quad \text{for every} \quad u \in V. \tag{2.2}$$

Proposition 2.1. It is the case that $B(u,v)$ is coercive over V. Moreover, $|B(u,v)| \leq c\|u\|_{1,\Omega}\|v\|_{1,\Omega}$ for $u,v \in V$, where c denotes a positive constant.

Proof. It is clear that

$$c_1 \sum_{r=1}^{2} \|D_r u\|^2 - c_3\|u\|^2 \leq B^\dagger(u,u) \leq c_2 \sum_{r=1}^{2} \|D_r u\|^2 + c_3\|u\|^2 \quad \text{for} \quad u \in V,$$

where the c_j denote positive constants. Moreover, it is well known (cf. [**73**], [**124**, p.49]) that for any $\epsilon > 0$ there exists the constant $c_\epsilon > 0$ such that

$$\int_\Gamma |tru|^2 ds \leq \epsilon \sum_{r=1}^{2} \|D_r u\|^2 + c_\epsilon\|u\|^2 \tag{2.3}$$

for every $u \in H^1(\Omega)$. Hence the first assertion of the theorem follows immediately from these results. The second assertion can similarly be proved if we make use of the Schwarz inequality.

Since $B(u,v)$ is coercive over V, it follows that B is bounded from below, that is, the infimum of $B(u,u)$ over the set $\{u \mid u \in V, \|u\| = 1\}$ is finite. We henceforth denote this infimum by γ and call γ the lower bound of B [**120**, p.310]. Let us now consider V only as a vector space, equip it with the inner product $(u,v)_B = B(u,v) - (\gamma - 1)(u,v)$, and let $\|u\|_B = (u,u)_B^{1/2}$. Then

Proposition 2.2. $\| \;\|_B$ and $\| \;\|_{1,\Omega}$ are equivalent norms on V, and hence V is a Hilbert space with respect to the inner product $(\;,\;)_B$.

Proof. Before beginning the proof let us remark that a particular consequence of the theorem is that the form B is closed [**120**, Theorem 11.1, p.314]. Turning now to the proof, we already know that V is a Hilbert space with respect to the inner product $(\;,\;)_{1,\Omega}$, and hence the theorem will be established once we have shown that $\| \;\|_{1,\Omega}$ and $\| \;\|_B$ are equivalent norms on V. Moreover, in light of Proposition 2.1, we need only show that there exists a constant $c > 0$ such that $\|u\|_{1,\Omega} \le c\|u\|_B$ for every $u \in V$. Accordingly, suppose that no such constant exists. Then there exists the sequence $\{u_n\}$ in V such that $\|u_n\|_{1,\Omega} = 1$ and $\|u_n\|_B \to 0$, and hence $u_n \to 0$ in $\mathcal{H}$. On the other hand, it follows from (2.2) that $\|u_n\|_B^2 \ge c_0 - (\gamma + c_1 - 1)\|u_n\|^2$, and so we arrive at the contradiction that $c_0 \le 0$.

We henceforth denote by A the selfadjoint operator in $\mathcal{H}$ associated with the form B. We note from [**120**, Theorem 2.6, p.323] that $A \ge \gamma$, $B(u,v) = (Au,v)$ for $u \in D(A)$ and $v \in V$, and that $D(A)$ is a core of B, that is, a dense subspace of V with respect to the norm $\| \;\|_{1,\Omega}$. Moreover, if $\mu \in \rho(A)$, then we know from [**4**, Lemma 3.4, p.210] that $(A - \mu I)^{-1}$ is bounded linear transformation from $\mathcal{H}$ into V, and hence it follows from Rellich's theorem [**4**, p.30] that A has compact resolvent. Thus we conclude from [**120**, p.187] that A has a discrete spectrum, so that in particular (see [**120**, pp.242-243]) $R(A - \mu I)$ is a closed subspace of $\mathcal{H}$ and $\dim N(A - \mu I) < \infty$ for every $\mu \in \mathbb{C}$.

We have derived the boundary value problem (2.1) in a formal way and we have stated above that this problem would be dealt with from a variational point of view. We are now in a position to make precise the definition of this problem. Accordingly, we define our problem as follows: determine pairs $\{\mu, u\}$, where $\mu \in \mathbb{C}$ and $u \in V$, for which

$$B(u,v) = \mu(\omega u, v) \quad \text{for every} \quad v \in V. \tag{2.4}$$

However, on account of the relationship between A and B cited above, the problem (2.4) may be formulated from a purely operator-theoretic point of view, and if we let T denote the bounded selfajoint operator in $\mathcal{H}$ defined by $(Tf)(x) = \omega(x)f(x)$, then this leads to the following definitions. We call μ an eigenvalue of the problem (2.1) if $Au = \mu Tu$ for some $u \ne 0$ in $D(A)$; u is called an eigenvector of (2.1) corresponding to μ. Now suppose that μ is an eigenvalue of (2.1) and let N_μ denote the set of all eigenvectors of (2.1) corresponding to μ together with the zero vector in $\mathcal{H}$. Then N_μ is a subspace of $\mathcal{H}$ which is called the eigenspace of (2.1) corresponding to μ and $\dim N_\mu$ is called the geometric multiplicity of μ. If $0 \neq u_0 \in N_\mu$, then it may happen that there exist the vectors $\{u_j\}_1^r$ in $D(A)$ such that $(A-\mu T)u_j = Tu_{j-1}$ for $j = 1, \ldots, r$. Then the vectors $\{u_j\}_1^r$ are said to be associated with the eigenvector u_0 and the set M_μ consisting of all eigenvectors of (2.1) corresponding to μ together with their associated vectors and the zero vector in $\mathcal{H}$ is a subspace of $\mathcal{H}$ which we call the principal subspace of (2.1) corresponding to μ and $\dim M_\mu$ is called the algebraic multiplicity of μ. If $N_\mu = M_\mu$, then we say that μ is semi-simple.

Although the problem (2.1) has only been presented in a formal way (that is, without worrying about the solution spaces), we shall continue to refer to this problem as the elliptic boundary value problem under investigation, but always bearing in mind that it is to be interpreted in the sense of equation (2.4), or equivalently, in the operator-theoretic sense described above.

For later use we require some further information concerning the operators A and T. Accordingly, if $\lambda \in \mathbb{C}$, then it follows from [**120**, Theorem 4.3, p.287] that $A - \lambda T$ is selfadjoint if λ is real. Now suppose that λ is not real and in $\mathcal{H}$ let us introduce the sesquilinear form $B_1(u,v) = (-\lambda Tu, v)$. Since B_1 is B-bounded with B-bound less than 1 [**120**, p.319], we conclude from [**120**, Theorem 1.33, p.320] that $\mathfrak{b} = B + B_1$ is sectorial and closed, while it follows easily from [**120**, Theorem 2.1, p.322] that the m-sectorial operator associated with $\mathfrak{b}$ is precisely $A - \lambda T$. Moreover, if $\mathfrak{b}^\star$ and $B_1^\star$ denote the adjoint forms of $\mathfrak{b}$ and B_1, respectively, in $\mathcal{H}$ [**120**, p.309], then it is a simple matter to verify that $B_1^\star(u,v) = (-\overline{\lambda}Tu, v)$, $\mathfrak{b}^\star = B + B_1^\star$, and that the m-sectorial operator associated with $\mathfrak{b}^\star$ is precisely $A - \overline{\lambda}T$. Hence it follows from [**120**, Theorem 2.5, p.323] that $A - \overline{\lambda}T$ is precisely the adjoint of $A - \lambda T$ in $\mathcal{H}$.

Finally, before we can proceed further with our investigation of the problem (2.1), we shall need some regularity results concerning solutions of (2.4); and this will be the subject of the following section.

2.4 Density and regularity results

In §2.3 we introduced the subspace V of $H^1(\Omega)$. However, from the point of view of integration by parts, the definition given there is not suitable, and hence it will be convenient to characterize V as the closure in $H^1(\Omega)$ of a certain class of smooth functions.

Definition 2.3. We let $C^\dagger(\Omega)$ denote: (i) $C_0^\infty(\Omega)$ if we are dealing with BVP I, (ii) $C_0^\infty(\mathbb{R}^2)|\Omega$ if we are dealing with BVP II, and (iii) the restriction to Ω of the class of functions $\phi \in C_0^\infty(\mathbb{R}^2)$ for which $\overline{\Gamma''} \cap \operatorname{supp} \phi = \emptyset$ if we are dealing with BVP III, where $^-$ denotes closure and supp = support.

It is clear that $C^\dagger(\Omega)$ may be identified with a subspace of $H^1(\Omega)$ and as such it is also a subspace of V. Moreover,

Theorem 2.1. The closure of $C^\dagger(\Omega)$ in $H^1(\Omega)$ is precisely V.

Proof. The theorem is known for the case of BVP II [**4**, Theorem 2.1, p.11]. Hence we shall only prove the theorem under the supposition that we are dealing with BVP III since the proof for the case of BVP I is similar. Furthermore, to simplify our work, we shall assume that $\alpha_r \neq 0$ and $\beta_r = \pi$ for $r = 1, 2$; the other cases can be similarly treated.

Accordingly, let $f \in V$ and let ϵ be any positive number. Then we have to show that there is a $\psi \in C^\dagger(\Omega)$ such that $\|f - \psi\|_{1,\Omega} < \epsilon$. To this end let n be an integer exceeding 1 and let $\{x^{(r)}\}_0^{4n-1}$ denote the points of Γ :

$$
\begin{aligned}
x^{(r)} &= (r/n, 0) \quad \text{for } r = 0, \ldots, n, \\
&= \big(1, (r-n)/n\big) \quad \text{for } r = (n+1), \ldots, 2n, \\
&= \big(1 - (r-2n)/n, 1\big) \quad \text{for } r = (2n+1), \ldots, 3n \\
&= \big(0, 1 - (r-3n)/n\big) \quad \text{for } r = (3n+1), \ldots, (4n-1).
\end{aligned}
$$

For $0 \leq r \leq 4n-1$, let O_r denote the disc $|x - x^{(r)}| < 1/n$ and Q_r the intersection of the square $|x_j - x_j^{(r)}| \leq 1/2n$, $j = 1, 2$, with I^2, where $x^{(r)} = (x_1^{(r)}, x_2^{(r)})$. Also let O_{4n} and Q_{4n} denote the squares defined by the inequalities $1/4n < x_j < 1 - 1/4n$, $j = 1, 2$, and $1/2n \leq x_j \leq 1 - 1/2n$, $j = 1, 2$, respectively. Then there exists the partition of unity $\{\varsigma_r\}_0^{4n}$ subordinate to the open covering $\{O_r\}_0^{4n}$ of I^2 such that for each r, $0 \leq \varsigma_r(x) \leq 1$ in $\mathbb{R}^2$, $\varsigma_r(x) = 1$ for $x \in Q_r$, $\operatorname{supp} \varsigma_r \subset O_r$ with $\operatorname{supp} \varsigma_r \subset \big\{x | x \in O_r,\ |x - x^{(r)}| \leq 11/12n\big\}$ for $r = 0, \ldots, (4n-1)$, and

$$|D_j \varsigma_r(x)| \leq cn \tag{2.5}$$

for $x \in \mathbb{R}^2$ and $j = 1, 2$, where c denotes a positive constant not depending upon j, r, or n. We henceforth let $\Omega_r = \Omega \cap O_r$, $f_r = f\varsigma_r$ for $r = 0, \ldots, 4n$ and assume that n is large enough so that

$$\|f\|_{1,\Omega_r} < \epsilon/6(2c + 2^{1/2}) \tag{2.6}$$

for $r = n$, $2n$ and $3n$.

Let us firstly show that $\|f_r\|_{1,\Omega} < \epsilon/6$ for $r = n$, $2n$, and $3n$. Accordingly, fixing our attention upon f_n, it follows from (2.5) that $\|f_n\|_{1,\Omega} \leq 2^{1/2}\|f\|_{1,\Omega_n} + 2cn\|f\|_{0,\Omega_n}$. On the other hand we know from [**134**, Theorem 2.7, p.73 and Lemma 3.5, p.178] that we may modify f on a set of measure zero if necessary and define $f(x) = 0$ for $x \in \Gamma_3$ so that for almost every x_2 in (0,1), $f(x_1, x_2)$ is absolutely continuous in $1/2 \leq x_1 \leq 1$ and $\partial f(x_1, x_2)/\partial x_1$ is equal to the distributional derivative of f with respect to x_1 almost everywhere in $1/2 < x_1 < 1$. Using this fact and arguing with Fubini's theorem, it is a simple matter to show that $\|f\|_{0,\Omega_n} \leq n^{-1}\|f\|_{1,\Omega_n}$, and hence it follows from (2.6) that $\|f_n\|_{1,\Omega} < \epsilon/6$. Similar arguments prove the same inequality with n replaced by $2n$ or $3n$.

Let us next fix our attention upon the remaining f_r. Since $\operatorname{supp} f_{4n} \subset O_{4n}$, we know (see [**4**, pp.5-7]) that there exists a $\psi_{4n} \in C_0^\infty(\mathbb{R}^2)$ such that $\operatorname{supp}\psi \subset O_{4n}$ and $\|f_{4n} - \psi_{4n}\|_{1,\Omega} < \epsilon/8n$. If $n < r < 2n$, then let us extend f_r to $\mathbb{R}^2$ by defining f_r to be zero in $\mathbb{R}^2\backslash\Omega_r$. Consequently it is well known that $f_r \in H^1(\mathbb{R}^2)$ (cf. [**134**, pp.184-185]). Hence if $\tau > 0$ is sufficiently small and $f_r^\tau(x) = f_r(x_1 + \tau, x_2)$, then $f_r^\tau \in H^1(\mathbb{R}^2)$, $\operatorname{supp} f_r^\tau \subset \Omega_r$, and $\|f_r^\tau - f_r\|_{1,\Omega} < \epsilon/16n$. Thus there exists a $\psi_r \in C_0^\infty(\mathbb{R}^2)$ such that $\operatorname{supp}\psi_r \subset \Omega_r$, $\|f_r^\tau - \psi_r\|_{1,\Omega} < \epsilon/16n$, and hence

$$\|f_r - \psi_r\|_{1,\Omega} < \epsilon/8n. \tag{2.7}$$

Similar arguments show that (2.7) also holds for $2n < r < 3n$. Turning to the case $0 < r < n$, let us extend f_r to $\mathbb{R}^2$ by defining f_r to be zero in $\overline{\mathbb{R}^2_+}\backslash\Omega_r$ and putting $f_r(x_1, x_2) = f_r(x_1, -x_2)$ for $x_2 < 0$, where $\mathbb{R}^2_+ = \{x | x = (x_1, x_2) \in \mathbb{R}^2,\ x_2 > 0\}$. Then (cf. [**134**, Proposition 3.4, p.185]) $f_r \in H^1(\mathbb{R}^2)$ and $\operatorname{supp} f_r \subset O_r$. Hence we may choose $\psi_r \in C_0^\infty(\mathbb{R}^2)$ such that $\operatorname{supp}\psi_r \subset O_r$ and for which (2.7) holds. Similar arguments show that (2.7) also holds for $3n < r < 4n$. Lastly to deal with the case $r = 0$, we extend f_0 to $\mathbb{R}^2$ by defining f_0 to be zero in $\mathbb{R}^2\backslash\Omega_0$. We also let O' denote the disc $|x| < 23/24n$ and put $\Gamma^\star = \overline{O'} \cap \Gamma$. Then we know from the proof of Theorem 2.1 of [**4**, pp.11-14] that $f_0 \in H^1(\mathbb{R}^2\backslash\Gamma^\star)$. Hence if we put $\Gamma_t^\star = \Gamma^\star - t\xi$, where ξ denotes the vector in $\mathbb{R}^2$ whose components are both $2^{-1/2}$ and $0 < t < 1/24n$, and if we let $f_0^t(x) = f_0(x + t\xi)$, then $f_0^t \in H^1(\mathbb{R}^2 \backslash \Gamma_t^*)$. Since we can choose t small enough so that $\|f_0^t - f_0\|_{1,\Omega} < \epsilon/16n$ and since we can find a function $\psi_0 \in C_0^\infty(\mathbb{R}^2)$ such that $\operatorname{supp} \psi_0 \subset O_0$ and $\|f_0^t - \psi_0\|_{1,\Omega} < \epsilon/16n$, we conclude that (2.7) also holds with $r = 0$.

Observing that $f(x) = \sum_{r=0}^{4n} f_r(x)$ for $x \in \Omega$, we see that the required function ψ is obtained by putting $\psi(x) = \sum' \psi_r(x)$, where $'$ indicates that the summation runs from $r = 0$ to $r = 4n$, but omits the indices n, $2n$ and $3n$.

Turning next to the regularity of solutions of (2.4), we associate with (2.4) the variational problem defined as follows: given $f \in \mathcal{H}$ find an element $u \in V$ such that

$$B(u, v) = (f, v) \quad \text{for every} \quad v \in V. \tag{2.8}$$

A vector u satisfying (2.8) will be called a solution of (2.8) corresponding to f.

Theorem 2.2. Let u be a solution of (2.8) corresponding to f. Then $u \in H^2(\Omega)$, u satisfies the boundary condition (2.1b) in the sense of trace, and $Lu = f$ in $\mathcal{H}$.

The proof of the theorem will depend upon the following lemma.

Lemma 2.1. Let u be a solution of (2.8) corresponding to f. Then $u \in H^2(G)$ for every open set $G \subset \Omega$ for which $\overline{G} \cap (\Gamma \backslash \cup_{r=1}^4 \Gamma_r) = \emptyset$ and $Lu = f$ almost everywhere in Ω. Moreover, if for $r = 1, \ldots, 4$ we let G_r be any relatively open subset of Γ_r whose relative closure is also contained in Γ_r, then u satisfies (2.1b), with Γ_r replaced by G_r, in the sense of trace.

Proof. For $x^0 \in \Omega$ and $0 < R < dist\{x^0, \Gamma\}$, let

$$S_R = \left\{ x | x \in \Omega, \quad |x - x^0| < R \right\}, \tag{2.9}$$

while for $x^0 \in \Gamma_r$ and $0 < R < dist\{x^0, \Gamma \backslash \Gamma_r\}$, let

$$\Sigma_R = \left\{ x | x \in \Omega, \quad |x - x^0| < R \right\}. \tag{2.10}$$

Then we know from [**4**, Theorem 9.2, p.107 and pp.111-112] that $u \in H^2(S_R)$ and that $u \in H^2(\Sigma_R)$ if we are dealing either with BVP I or with BVP III and $\Gamma_r \subset \Gamma''$. Thus in particular $u \in H^2_{loc}(\Omega)$, and hence if $\phi \in C_0^\infty(\Omega)$, then in the expression $B^\dagger(u,\phi)$ we may transfer all the differentiation from ϕ onto u and obtain

$$(f,\phi) = B(u,\phi) = (Lu,\phi).$$

Thus we conclude that $Lu = f$ almost everywhere in Ω.

Let us suppose that we are dealing either with BVP II or with BVP III and $\Gamma_r \subset \Gamma'$ and show that if $0 < R' < R$, then $u \in H^2(\Sigma_{R'})$. Moreover, in order to simplify the argument, we will also suppose that $r = 2$. Accordingly, let $R'' = (R + R')/2$, let $\varsigma \in C^\infty(\mathbb{R}^2)$ such that $\varsigma = 1$ for $|x - x^0| < R'$, $\varsigma = 0$ for $|x - x^0| > R''$, and for $h \in \mathbb{R}$, $0 < |h| < R - R''$, let $v_h(x) = (\delta_h \varsigma u)(x)$, where

$$(\delta_h g)(x) = \Big(g(x_1 + h, x_2) - g(x_1, x_2)\Big)/h. \tag{2.11}$$

It is important to note that $v_h \in V$. If we now argue with the expression $B(v_h,\phi)$, $\phi \in C^\dagger(\Omega)$, as in the proof of Theorem 9.2 of [**4**, p.107] and make use of (2.3), then we may show that

$$\Big|B(v_h,\phi)\Big| \le \Big|B(u,\varsigma\delta_{-h}\phi)\Big| + c_1\|\phi\|_{1,\Sigma_R}\|u\|_{1,\Sigma_R},$$

where the constant c_1 does not depend upon h or ϕ. Hence, since $\varsigma\delta_{-h}\phi \in V$, it follows from (2.8) and [**4**, Theorem 3.13, p.42] that

$$\begin{aligned}\Big|B(v_h,\phi)\Big| &\le \|f\|\,\|\varsigma\delta_{-h}\phi\| + c_1\|\phi\|_{1,\Sigma_R}\|u\|_{1,\Sigma_R} \\ &\le c_2\Big(\|f\| + \|u\|_{1,\Sigma_R}\Big)\|\phi\|_{1,\Sigma_R},\end{aligned}$$

where the constant c_2 does not depend upon h or ϕ. Thus we conclude from the fact that $C^\dagger(\Omega)$ is dense in V that

$$\Big|B(v_h,v_h)\Big| \le c_2\Big(\|f\| + \|u\|_{1,\Sigma_R}\Big)\|v_h\|_{1,\Sigma_R},$$

and hence, in light of (2.2),

$$\|v_h\|_{1,\Sigma_R} \le c_3\Big(\|f\| + \|u\|_{1,\Sigma_R}\Big),$$

where the constant c_3 does not depend upon h. It follows immediately from [**4**, Theorem 3.16, p.45] that $D_1 u \in H^1(\Sigma_{R'})$. On the other hand, since

$$a_2 D_2^2 u = -(D_2 a_2) D_2 u - D_1 a_1 D_1 u + qu - f \text{ almost everywhere in } \Sigma_{R'},$$

it also follows that $D_2 u \in H^1(\Sigma_{R'})$, and hence $u \in H^2(\Sigma_{R'})$. It is now obvious that $u \in H^2(G)$ for every open set $G \subset \Omega$ for which $\overline{G} \cap (\Gamma \setminus \cup_{r=1}^4 \Gamma_r) = \emptyset$, and hence the proof of the first assertion of the lemma is complete.

To prove the second assertion let us suppose that we are dealing either with BVP II or BVP III and $\Gamma_2 \subset \Gamma'$, that $x^0 = (x_1^0, x_2^0) \in \Gamma_2$, and $0 < R < \min\{x_1^0, 1 - x_1^0\}$. For $r = 1, 2$ let $\psi_r(x_r) \in C^\infty(\mathbb{R})$ such that $\operatorname{supp} \psi_1 \subset \{x_1 \mid |x_1 - x_1^0| < R\}$, $\psi_2(x_2) = 1$ for $|x_2| < 1/4$, $\psi_2(x_2) = 0$ for $|x_2| > 1/2$, and let $\psi(x) = \prod_{r=1}^2 \psi_r(x_r)$ (note that $\psi|\Omega \in C^\dagger(\Omega)$). Hence if $\phi \in C_0^\infty(\mathbb{R}^2)$, then integration by parts show that

$$B^\dagger(\phi, \psi) = (L\phi, \psi) + \int_{\Gamma_2} b_2 (\partial \phi / \partial \nu_2) \overline{\psi_1} ds,$$

and since $u \in H^2(\Omega')$, where $\Omega' = \{x | x \in \Omega, \ |x_1 - x_1^0| < R\}$, we conclude from [**4**, Theorem 2.1, p.11] that

$$B(u, \psi) = (Lu, \psi) + \int_{\Gamma_2} \left[b_2 tr \partial u / \partial \nu_2 + \sigma_2 tr u \right] \overline{\psi_1} ds.$$

In light of (2.8) and the facts that $Lu = f$ in $\mathcal{H}$ and ψ_1 is arbitrary, we see that the second assertion of the lemma is true for the case under consideration when $r = 2$ and $G_2 = \{x \in \mathbb{R}^2 \mid |x_1 - x_1^0| < R, \ x_2 = 0\}$. We complete the proof of the lemma by arguing in the same way on other portions of Γ if we are dealing with BVP II or BVP III and by appealing to the definition of V.

Proof of Theorem 2.2. In light of Lemma 2.1 it remains only to prove that $u \in H^2(\Omega)$; and to this end we are going to consider extensions of u to larger regions containing Ω which are obtained by reflecting about the boundary Γ. Accordingly, let I_1^2, denote the rectangle defined by the inequalities $-1 \le x_1 \le 1$, $0 \le x_2 \le 1$, let Ω_1 denote the interior of I_1^2, and let

$$\Gamma_{11} = \left\{ x \in \mathbb{R}^2 | x_1 = -1, \ 0 < x_2 < 1 \right\},$$

$$\Gamma_{12} = \left\{ x \in \mathbb{R}^2 | -1 < x_1 < 1,\ x_2 = 0 \right\},$$

$$\Gamma_{13} = \left\{ x \in \mathbb{R}^2 | x_1 = 1,\ 0 < x_2 < 1 \right\},$$

$$\Gamma_{14} = \left\{ x \in \mathbb{R}^2 | -1 < x_1 < 1,\ x_2 = 1 \right\},$$

denote the faces of Ω_1. Moreover, if we are dealing either with BVP II or with BVP III, then let Γ_1'' denote the union of Γ_{11}, Γ_{13}, and, if we are dealing with BVP III, those Γ_{1r}, $r = 2$ or 4, for which $\Gamma_r \subset \Gamma''$ (if there are any), and let $\Gamma_1' = \partial\Omega_1 \backslash \Gamma_1''$ (∂ = boundary). Let T_1 denote the transformation in $\mathbb{R}^2 : (x_1, x_2) \to (-x_1, x_2)$. Then if we are dealing either with BVP II or with BVP III and $\Gamma_1'' \neq \cup_{r=1}^4 \Gamma_{1r}$, and if $\Gamma_{1r} \subset \Gamma_1'$, we define the function $\sigma_r^\dagger$ on Γ_{1r} by putting

$$\sigma_r^\dagger(x) = \sigma_r(x) \text{ on that portion of } \Gamma_{1r} \text{ for which } x_1 \geq 0,$$

$$= \sigma_r(T_1 x) \text{ on that portion of } \Gamma_{1r} \text{ for which } x_1 < 0.$$

If we are dealing with BVP I (resp. BVP II or BVP III), then let V_1 denote the subspace of $H^1(\Omega_1)$ consisting of those elements which vanish, in the sense of trace, on all the Γ_{1r} (resp. on those Γ_{1r} for which $\Gamma_{1r} \subset \Gamma_1''$). It is clear that V_1 is a closed subspace of $H^1(\Omega_1)$. Now let $C^\dagger(\Omega_1)$ denote : (i) $C_0^\infty(\Omega_1)$ if we are dealing either with BVP I or with BVP III and $\Gamma_1'' = \cup_{r=1}^4 \Gamma_{1r}$ and (ii) the restriction to Ω_1 of the class of functions $\phi \in C_0^\infty(\mathbb{R}^2)$ for which $\overline{\Gamma_1''} \cap \operatorname{supp} \phi = \emptyset$ if we are dealing either with BVP II or with BVP III and $\Gamma_1'' \neq \cup_{r=1}^4 \Gamma_{1r}$. Then if we identify $C^\dagger(\Omega_1)$ with a subspace of V_1, we may argue as we did in the proof of Theorem 2.1 to show that $C^\dagger(\Omega_1)$ is dense in V_1.

Let us next fix our attention upon the vector u of the theorem and let $u_1^\# = \phi(x_1)u(x)$, where

$$\phi \in C^\infty(\mathbb{R}),\ 0 \leq \phi(x_1) \leq 1,\ \phi(x_1) = 1 \text{ for } x_1 \leq 1/4,\ \phi(x_1) = 0 \text{ for } x_1 \geq 3/4.$$

We now define $u_1(x)$ in Ω_1 in the following way:

(1) if we are dealing either with BVP I or with BVP III and $\Gamma_1 \subset \Gamma''$, then let

$$u_1(x) = u_1^\#(x) \text{ for } x_1 > 0,\ u_1(x) = -u_1(T_1 x) \text{ for } x_1 < 0;$$

(2) if we are dealing either with BVP II or with BVP III and $\Gamma_1 \subset \Gamma'$, then let

$$u_1(x) = \big(1 + \chi_1(x_1)\big)u_1^{\#}(x) \text{ for } x_1 > 0,\ u_1(x) = u_1(T_1 x) \text{ for } x_1 < 0,$$

where $\chi_1(x_1) = -x_1\rho_1(x_1)\big(p_1(0)\big)^{-1}\cot\alpha_1$, $\rho_1 \in C^\infty(\mathbb{R})$, $0 \le \rho_1(x_1) \le 1$, $\rho_1(x_1) = 1$ for $x_1 \le 0$, $\rho_1(x_1) = 0$ for $x_1 \ge \delta_1$, and $2\delta_1 = \min\{1, p_1(0)|\tan\alpha_1|\}$. It follows from the definition of u_1 and standard arguments (cf. [**134**, pp.184-185]) that $u_1 \in V_1$.

We are now going to extend the form B to V_1. Accordingly, for $x \in I_1^2$ we define $a_r^\dagger(x)$ and $q^\dagger(x)$ to be $a_r(x)$ and $q(x)$, respectively, if $x_1 \ge 0$ and to be $a_r(T_1 x)$ and $q(T_1 x)$, respectively, if $x_1 < 0$. Let

$$B_1^\dagger(w,v) = \sum_{r=1}^{2}(D_r w, a_r^\dagger D_r v)_{0,\Omega_1} + (w, q^\dagger v)_{0,\Omega_1} \text{ for } w,v \in H^1(\Omega_1).$$

If we are dealing either with BVP I or with BVP III and $\Gamma_1'' = \cup_{r=1}^4 \Gamma_{1r}$, then let $B_1(w,v) = B_1^\dagger(w,v)$ for $w,v \in V_1$. If we are dealing either with BVP II or with BVP III and $\Gamma_1'' \ne \cup_{r=1}^4 \Gamma_{1r}$, then let

$$B_1(w,v) = B_1^\dagger(w,v) + \int_{\Gamma_1'} \sigma^\dagger(trw)(\overline{trv})ds \text{ for } w,v \in V_1,$$

where $\sigma^\dagger(x) = \sigma_r^\dagger(x)$ if $x \in \Gamma_{1r}$ and $\Gamma_{1r} \subset \Gamma_1'$. We may argue as we did in the proof of Proposition 2.1 to show that $B_1(w,v)$ is coercive over V_1.

Let us now show that there is an $f_1 \in L^2(\Omega_1)$ such that

$$B_1(u_1,v) = (f_1,v)_{0,\Omega_1} \text{ for every } v \in V_1. \tag{2.12}$$

To this end let us suppose firstly that we are dealing either with BVP II or with BVP III and $\Gamma_1 \subset \Gamma'$ and let $\psi \in C^\dagger(\Omega_1)$. Then observing that

$$(D_1u_1)(x) = -(D_1u_1)(T_1x) \text{ and } (D_2u_1)(x) = (D_2u_1)(T_1x)$$
$$\text{for } x \in \Omega_1^- = \{x \in \Omega_1 \mid x_1 < 0\},$$

it follows that

$$B_1(u_1,\psi) = B(\varsigma_1 u, \psi^\dagger) - 2\int_{\Gamma_1}\sigma_1(tru)\overline{\psi}ds$$

$$= B(u, \varsigma_1\psi^\dagger) - 2\int_{\Gamma_1} \sigma_1(tru)\overline{\psi}ds + \tag{2.13}$$

$$+ \left(a_1(D_1\varsigma_1)u, D_1\psi^\dagger\right) - \left(a_1(D_1\varsigma_1)D_1u, \psi^\dagger\right),$$

where $\varsigma_1(x_1) = \big(1 + \chi_1(x_1)\big)\phi(x_1)$ and $\psi^\dagger(x) = \psi(x) + \psi(T_1x)$. On the other hand we know from [**134**, Theorem 2.7, p.73 and Lemma 3.5, p.178] that we may modify u on a set of measure zero and extend u to the boundary by continuity so that for almost every x_2 in $(0,1)$, $u(x_1, x_2)$ is absolutely continuous in $0 \le x_1 \le 1$ and $\partial u(x_1, x_2)/\partial x_1 = (D_1u)(x_1, x_2)$ almost everywhere in $0 < x_1 < 1$, where D_1u denotes the distributional derivative of u with respect to x_1. From this fact and Fubini's theorem it follows that

$$\left(a_1(D_1\varsigma_1)u, D_1\psi^\dagger\right) = 2\int_{\Gamma_1} \sigma_1(tru)\overline{\psi}ds - \left(D_1a_1(D_1\varsigma_1)u, \psi^\dagger\right),$$

and hence in light of (2.8), (2.13) and the fact that $\varsigma_1\psi^\dagger \in C^\dagger(\Omega)$ we conclude that

$$B_1(u_1, \psi) = (\varsigma_1 f + g_1, \psi^\dagger),$$

where

$$g_1 = -2a_1(D_1\varsigma_1)D_1u - u\left[(D_1a_1)(D_1\varsigma_1) + a_1D_1^2\varsigma_1\right].$$

Since $C^\dagger(\Omega_1)$ is dense in V_1, it follows immediately that for the case under consideration here (2.12) is valid with

$$f_1(x) = (\varsigma_1 f + g_1)(x) \quad \text{for} \quad x_1 > 0,$$

$$= f_1(T_1x) \quad \text{for} \quad x_1 < 0.$$

If we are dealing either with BVP I or with BVP III and $\Gamma_1 \subset \Gamma''$, then we may argue as above to show that (2.12) is again valid with

$$f_1(x) = (\phi f + g)(x) \quad \text{for} \quad x_1 > 0,$$

$$= -f_1(T_1x) \quad \text{for} \quad x_1 < 0,$$

where

$$g = -2a_1(D_1\phi)D_1u - u\left[(D_1a_1)(D_1\phi) + a_1D_1^2\phi\right].$$

In light of (2.12) we may now argue as we did in the proof of Lemma 2.1 to show that $u_1 \in H^2(G)$ for every open set $G \subset \Omega_1$ for which $\overline{G} \cap (\partial\Omega_1 \setminus \cup_{r=1}^4 \Gamma_{1r}) = \emptyset$. Since $\operatorname{supp} u_1 \cap (\overline{\Gamma}_{11} \cup \overline{\Gamma}_{13}) = \emptyset$, we conclude that for each $x \in I_1^2$, there is an open disc G_x in $\mathbb{R}^2$ with centre x such that $u_1 \in H^2(\Omega_1 \cap G_x)$, and hence it follows that $u_1 \in H^2(\Omega_1)$. Note also that (2.12) implies that

$$L_1u_1 = f_1 \text{ almost everywhere in } \Omega_1, \tag{2.14}$$

where L_1 denotes the elliptic operator $-\sum_{r=1}^2 D_r a_r^\dagger(x) D_r + q^\dagger(x)$.

Let us again fix our attention upon the vector u of the theorem and let $u_3^\#(x) = (1 - \phi(x_1))u(x)$. If Ω_3 denotes the region defined by the inequalities $0 < x_1 < 2$, $0 < x_2 < 1$, then we define u_3 in Ω_3 in the following way:

(1) if we are dealing either with BVP I or with BVP III and $\Gamma_3 \subset \Gamma''$, then let

$$u_3(x) = u_3^\#(x) \text{ for } x_1 < 1,\ u_3(x) = -u_3(T_3x) \text{ for } x_1 > 1,$$

where T_3 denotes the transformation in $\mathbb{R}^2 : (x_1, x_2) \to (2 - x_1, x_2)$;

(2) if we are dealing either with BVP II or with BVP III and $\Gamma_3 \subset \Gamma'$, then let

$$u_3(x) = (1 + \chi_3(x_1))u_3^\#(x) \text{ if } x_1 < 1,\ u_3(x) = u_3(T_3x) \text{ for } x_1 > 1,$$

where $\chi_3(x_1) = -(x_1 - 1)\rho_3(x_1)(p_1(1))^{-1}\cot\beta_1$, $\rho_3 \in C^\infty(\mathbb{R})$, $0 \le \rho_3(x_1) \le 1$, $\rho_3(x_1) = 1$ for $x_1 \ge 1$, $\rho_3(x_1) = 0$ for $x_1 \le 1 - \delta_3$, and $2\delta_3 = \min\{1, p_1(1)|\tan\beta_1|\}$. We can now argue with u_3 as we argued with u_1 above to show that $u_3 \in H^2(\Omega_3)$.

Finally, since $u_1 + u_3 \in H^2(\Omega)$ and $u_1 + u_3 = (1 + \chi)u$ in $\mathcal{Y}$, where $\chi \in C^2(I^2)$ and $|\chi(x)| \le 1/2$ for $x \in I^2$, we conclude that $u \in H^2(\Omega)$, which completes the proof of the theorem.

<u>Theorem 2.3.</u> Suppose that $2 < p < \infty$, $f \in L^p(\Omega)$ and u is a solution of (2.8) corresponding to f. Then $u \in W^{2,p}(\Omega)$.

In order to prove the theorem we require some preliminary results. Accordingly, we henceforth let $\|\ \ \|_{k,p,G}$ denote the norm in $W^{k,p}(G)$ and for $x^0 = (x_1^0, x_2^0) \in I^2$ let L^0

denote the principal part of L with coefficients frozen at their values at x^0. Then fixing our attention upon the case $x^0 \in \Gamma_2$ and recalling the definition of Σ_R given by (2.10) (taking $r = 2$ now), we have

Lemma 2.2. If $g \in C_0^\infty(\Sigma_R)$, then there exists the function $z \in C^\infty(\overline{\Sigma}_R)$ such that

$$L^0 z = g \text{ in } \Sigma_R,$$

$$D_2 z = 0 \text{ on } |x_1 - x_1^0| < R,\ x_2 = 0,$$

$$\|z\|_{2,p',\Sigma_R} \leq c\|g\|_{0,p',\Sigma_R},$$

where $1/p' = 1 - 1/p$ and the constant c depends only upon p', R, and the a_j.

Proof. It is well known (cf. [**2**], [**113**]) that there is a fundamental solution $F(x)$ of L^0 such that if we put $g(x) = 0$ for $x \in \mathbb{R}^2 \backslash \Sigma_R$ and let

$$v(x) = \int_{\mathbb{R}^2} F(x-y)g(y)dy \text{ for } x \in \mathbb{R}^2,$$

then

$$v \in C^\infty(\mathbb{R}^2),\ L^0 v = g \text{ in } \mathbb{R}^2, \text{ and } \|v\|_{2,p',S_{2R}} \leq c_1\|g\|_{0,p',\Sigma_R}, \tag{2.15}$$

where S_{2R} denotes the disc $|x - x^0| < 2R$ and the constant c_1 depends only upon p', R, and the a_j.

Let $\varsigma \in C^\infty(\mathbb{R}^2)$ with $\varsigma(x) = 1$ for $|x - x^0| < R$ and $\operatorname{supp}\varsigma \subset S_{2R}$, and let us consider the boundary value problem

$$\begin{aligned} &L^0 w = 0 \text{ in } \mathbb{R}^2_+, \\ &D_2 w = D_2 v^\dagger \text{ on } x_2 = 0, \end{aligned} \tag{2.16}$$

where $\mathbb{R}^2_+ = \{x = (x_1, x_2) \in \mathbb{R}^2 | x_2 > 0\}$ and $v^\dagger = \varsigma v$. We know from [**5**, Theorem 2.1] that there exists the kernel $K(x_1, x_2)$ defined and infinitely differentiable in $\overline{\mathbb{R}^2_+} \backslash \{0\}$ such that a $C^\infty(\overline{\mathbb{R}^2_+})$ solution of (2.16) is given by

$$w(x) = \int_{\mathbb{R}} K(x_1 - y_1, x_2) D_2 v^\dagger(y_1, 0)\, dy_1. \tag{2.17}$$

We also know from the reference just cited that if $r \in \mathbb{N}$ is odd, then K admits the representation

$$K(x_1, x_2) = D_1^{1+r} K_r(x_1, x_2),$$

where K_r is infinitely differentiable in $\overline{\mathbb{R}^2_+}\backslash\{0\}$,

$$\begin{aligned} \left|D^s K_r(x)\right| &\le c_1 |x|^{r+1-|s|}\left(1 + \left|\log|x|\right|\right) \text{ if } |s| \le r+1, \\ &\le c_2 |x|^{r+1-|s|} \text{ if } |s| > r+1, \end{aligned} \tag{2.18}$$

s denotes the multi-index $(s_1, s_2), |s| = s_1 + s_2$, $D^s = D_1^{s_1} D_2^{s_2}$, and the constant c_2 depends only upon r, s, and the a_j. Hence if we observe from (2.17) that

$$\begin{aligned} w(x) = &- \int_{\mathbb{R}^2_+} D_2 K(x_1 - y_1, x_2 + y_2) D_2 v^\dagger(y_1, y_2)\, dy - \\ &- \int_{\mathbb{R}^2_+} K(x_1 - y_1, x_2 + y_2) D_2^2 v^\dagger(y_1, y_2)\, dy \text{ for } x \in \mathbb{R}^2_+, \end{aligned}$$

then it follows that

$$\begin{aligned} D^s w(x) = &\int_{\mathbb{R}^2_+} D_2 D_1^r D^s K_r(x_1 - y_1, x_2 + y_2) D_1 D_2 v^\dagger(y_1, y_2)\, dy - \\ &- \int_{\mathbb{R}^2_+} D_1^{1+r} D^s K_r(x_1 - y_1, x_2 + y_2) D_2^2 v^\dagger(y_1, y_2)\, dy \text{ for } x \in \mathbb{R}^2_+. \end{aligned}$$

Taking now $r = 1$, we conclude from (2.18) and [**2**, Sublemma] that

$$\|D^s w\|_{0,p',\mathbb{R}^2_+} \le c_3 \|v^\dagger\|_{2,p',S_{2R}} \text{ if } |s| = 2 \tag{2.19}$$

while (2.18) and standard potential-theoretic arguments show that

$$\|D^s w\|_{0,p',\Sigma_R} \le c_4 \|v^\dagger\|_{2,p',S_{2R}} \text{ if } |s| < 2, \tag{2.20}$$

where the c_i depend only upon p', the a_j, and, in the case of c_4, also upon R. Hence if we let $z = v^\dagger - w$, then it follows from (2.15-16) and (2.19-20) that z satisfies all the assertions of the lemma.

Lemma 2.3. Suppose that $2 < p < \infty$, $f \in L^p(\Omega)$, and u is a solution of (2.8) corresponding to f. Then $u \in W^{2,p}(G)$ for every open set $G \subset \Omega$ for which $\overline{G} \cap (\Gamma \backslash \cup_{r=1}^4 \Gamma_r) = \emptyset$.

Proof. To begin with, we note from the Sobolev imbedding theorem [**1**, p.97] that $u \in H^2(\Omega)$ implies that $u \in W^{1,p}(\Omega)$. Now suppose firstly that $x^0 \in \Omega$, that S_R is the open disc (2.9), and that $\phi \in C_0^\infty(S_R)$. If we put $\phi(x) = 0$ for $x \in \Omega \backslash S_R$, then it follows from an approximation of u by smooth functions [**4**, Theorem 2.1, p.11], integration by parts, and going to the limit, that for $1 \le i \le 2$,

$$(D_i u, L\phi) = -B(u, D_i\phi) + (u, qD_i\phi) - \sum_{r=1}^{2}(D_r u, (D_i a_r)D_r\phi) + (D_i u, q\phi). \quad (2.21)$$

Hence, in light of (2.8), we conclude that

$$|(D_i u, L\phi)| \le c_1 \left[\|f\|_{0,p,S_R} + \|u\|_{1,p,S_R} \right] \|\phi\|_{1,p',S_R},$$

where the constant c_1 does not depend upon ϕ. It follows immediately from [**2**, Lemma 5.1] that $u \in W^{2,p}(S_{R'})$ for some $R' < R$.

Suppose next that $x^0 \in \Gamma_2$ and that we are dealing either with BVP I or with BVP III and $\Gamma_2 \subset \Gamma''$. Suppose also that Σ_R is given by (2.10) (taking $r = 2$ now) and that $\phi \in C^\infty(\overline{\Sigma}_R)$, with $\phi = 0$ on that part of boundary of Σ_R situated on the line $x_2 = 0$ and $\phi = 0$ in a neighbourhood of that part of the boundary of Σ_R situated on the circle $|x - x^0| = R$. If we put $\phi(x) = 0$ for $x \in \Omega \backslash \overline{\Sigma}_R$ and argue as we did with the case $x^0 \in \Omega$, then it follows that (2.21) is again valid for $i = 1$. Hence, in light of (2.8), we conclude that

$$|(D_1 u, L\phi)| \le c_2 \left[\|f\|_{0,p,\Sigma_R} + \|u\|_{1,p,\Sigma_R} \right] \|\phi\|_{1,p',\Sigma_R}, \quad (2.22)$$

where the constant c_2 does not depend upon ϕ. It follows immediately from [**2**, Lemma 5.2] that $D_1^2 u \in L^p(\Sigma_{R'})$ for some $R' < R$. Moreover, since $Lu = f$ almost everywhere in Ω, it also follows that $D_2^2 u \in L^p(\Sigma_{R'})$. Thus we have $D_1 u \in L^p(\Sigma_{R'})$, $D_1^2 u \in L^p(\Sigma_{R'})$, and $|(D_1 u, D_2^2\phi)_{\Sigma_{R'}}| = |(D_2^2 u, D_1\phi)_{\Sigma_{R'}}| \le \|D_2^2 u\|_{0,p,\Sigma_{R'}} \|\phi\|_{1,p',\Sigma_{R'}}$ for every

$\phi \in C_0^\infty(\Sigma_{R'})$, where for any open set $G \subset \mathbb{R}^2$, $(\ , \ )_G$ denotes the inner product in $L^2(G)$, and hence follows from [**2**, Lemma 6.1] that $u \in W^{2,p}(\Sigma_{R''})$ for every $R'' < R'$.

Finally, let us suppose that $x^0 \in \Gamma_2$ and that we are dealing either with BVP II or with BVP III and $\Gamma_2 \subset \Gamma'$. Suppose also that Σ_R is given by (2.10) (with $r = 2$ now) and that $\phi \in C^\infty(\overline{\Sigma}_R)$, with $D_2\phi = 0$ on that part of the boundary of Σ_R situated on the line $x_2 = 0$ and $\phi = 0$ in a neighbourhood of that part of the boundary of Σ_R situated on the circle $|x - x^0| = R$. Then we may argue precisely as above to show that (2.22) is valid for this case also. Note that (2.22) implies that

$$\left|(D_1 u, L'\phi)_{\Sigma_R}\right| \leq c_3 \|\phi\|_{1,p',\Sigma_R}, \tag{2.23}$$

where L' denotes the principal part of L and the constant c_3 depends upon u, f, p, R, the a_j, and q, but not upon ϕ.

For $0 < r \leq R/2$ and $j = 1,2$, let $\varsigma_{rj} \in C^\infty(\mathbb{R})$ such that $0 \leq \varsigma_{rj}(x_j) \leq 1$, $\varsigma_{rj}(x_j) = 1$ for $|x_j| < r/3$, $\varsigma_{rj}(x_j) = 0$ for $|x_j| > 2^{1/2}r/3$, and let $\varsigma_r(x) = \prod_{j=1}^2 \varsigma_{rj}(x_j)$. Then if $v \in C^\infty(\overline{\Sigma}_R)$, $D_2 v = 0$ on $x_2 = 0$, and if we put $\varsigma_r v = 0$ outside of $\overline{\Sigma}_r$, then it follows from (2.23) and the fact that

$$\left|(D_1 u, L'\varsigma_r v)_{\Sigma_r} - (\varsigma_r D_1 u, L'v)_{\Sigma_r}\right| \leq k_1(r)\|u\|_{1,p,\Sigma_r}\|v\|_{1,p',\Sigma_r},$$

that

$$\left|(u_r, L'v)_{\Sigma_r}\right| \leq k_2(r)\|v\|_{1,p',\Sigma_r}, \tag{2.24}$$

where $u_r = \varsigma_r D_1 u$ and the constants $k_j(r)$ do not depend upon v, but depend upon r. Now for $h \in \mathbb{R}$, $0 < |h| < r/6$, and δ_h defined by (2.11), let $g_h = |\delta_h u_r|^{p-1} sgn \delta_h u_r$, where $sgn\ g = g|g|^{-1}$ if $g \neq 0$ and is zero otherwise. Then $g_h \in L^{p'}(\Sigma_r)$ and

$$\|g_h\|_{0,p',\Sigma_r} = \|\delta_h u_r\|_{0,p,\Sigma_r}^{p-1}. \tag{2.25}$$

Choose $g_h^\dagger \in C_0^\infty(\Sigma_r)$ such that

$$\|g_h - g_h^\dagger\|_{0,p',\Sigma_r} \leq 3^{-1}\|g_h\|_{0,p',\Sigma_r}, \tag{2.26}$$

and taking $g = g_h^\dagger$ in Lemma 2.2, let us now denote the function z there by v_h. We then have

$$\|v_h\|_{2,p',\Sigma_R} \leq (4c/3)\|\delta_h u_r\|_{0,p,\Sigma_r}^{p-1}, \tag{2.27}$$

and hence it follows from (2.24) that

$$\begin{aligned} \left|(u_r, L'\delta_{-h}v_h)_{\Sigma_r}\right| &\leq k_2(r)\|\delta_{-h}v_h\|_{1,p',\Sigma_r} \leq k_2(r)\|v_h\|_{2,p',\Sigma_r} \\ &\leq (4c/3)k_2(r)\|\delta_h u_r\|_{0,p,\Sigma_r}^{p-1}. \end{aligned} \tag{2.28}$$

Moreover, since

$$(L'\delta_{-h}v_h)(x) = (\delta_{-h}L'v_h)(x) - \sum_{j=1}^{2}(\delta_{-h}a_j)(x)D_j^2 v_h(x - he_1),$$

where e_1 denotes the unit vector in $\mathbb{R}^2$ parallel to and in the direction of the positive x_1 - axis, we conclude from (2.27-28) that

$$\left|(\delta_h u_r, L'v_h)_{\Sigma_r}\right| \leq k_3(r)\|\delta_h u_r\|_{0,p,\Sigma_r}^{p-1}, \tag{2.29}$$

where the constant $k_3(r)$ depends upon r, but not upon h. Finally, since

$$\begin{aligned} \|\delta_h u_r\|_{0,p,\Sigma_r}^{p} &= (\delta_h u_r, g_h)_{\Sigma_r} \\ &= (\delta_h u_r, L'v_h)_{\Sigma_r} + (\delta_h u_r, (L^0 - L')v_h)_{\Sigma_r} + (\delta_h u_r, g_h - g_h^{\dagger})_{\Sigma_r}, \end{aligned}$$

and

$$\left|(\delta_h u_r, (L^0 - L')v_h)_{\Sigma_r}\right| \leq \kappa r\|\delta_h u_r\|_{0,p,\Sigma_r}\|v_h\|_{2,p',\Sigma_r},$$

where the constant κ depends only upon the a_j, it follows immediately from (2.25-27) and (2.29) that

$$\|\delta_h u_r\|_{0,p,\Sigma_r} \leq 3k_3(r)$$

provided that we take $r \leq \min\{R/2, 1/4c\kappa\}$. Choosing r accordingly, we conclude that $D_1^2 u \in L^p(\Sigma_{r/3})$, and hence we may now argue as we did before to show that $u \in$

$W^{2,p}(\Sigma_{r'})$ for every $r' < r/3$.

The assertion of the lemma now follows from the results.

Proof of Theorem 2.3. Let us fix our attention upon the proof of Theorem 2.2 and consider the extension u_1 of u to the region Ω_1. Then we know that $u_1 \in H^2(\Omega_1)$, that (2.12) and (2.14) are valid, and that $f_1 \in L^p(\Omega_1)$. Hence we may argue precisely as we did in the proof of Lemma 2.3 to show that $u_1 \in W^{2,p}(G)$ for every open set $G \subset \Omega_1$ for which $\overline{G} \cap (\partial\Omega_1 \setminus \cup_{r=1}^4 \Gamma_{1r}) = \emptyset$. But we know from the arguments given in the proof of Theorem 2.2 that this implies that $u_1 \in W^{2,p}(\Omega_1)$. Similarly, if we consider the extension u_3 of u to Ω_3, then we can show that $u_3 \in W^{2,p}(\Omega_3)$. Consequently, we may argue as we did in the last paragraph of the proof of Theorem 2.2 to show that $u \in W^{2,p}(\Omega)$, which completes the proof of the theorem.

2.5. The system (1.1-4)

We are now going to establish some relationships between the eigenvalues and eigenfunctions of the system (1.1-4) and the eigenvalues and eigenvectors of the problem (2.1). To this end we firstly need

Theorem 2.4. An element u in V belongs to $D(A)$ if and only if $u \in H^2(\Omega)$ and satisfies (2.1b) in the sense of trace. Moreover, $Au = Lu$ in $\mathcal{H}$ for $u \in D(A)$.

Proof. To begin with, let us fix our attention upon the first assertion, suppose that $u \in D(A)$, and let $Au = f$. Then we know that $B(u, v) = (f, v)$ for every $v \in V$, and hence the proof of one part of the assertion follows from the Theorem 2.2. To prove the other part, suppose that $\phi \in C_0^\infty(\mathbb{R}^2)$, $\psi \in C^\dagger(\Omega)$, and let

$$p(x) = p_r(0)A_{3-r}(x_{3-r}) \text{ if } x \in \Gamma_r \text{ and } 1 \le r \le 2,$$

$$= p_{r-2}(1)A_{5-r}(x_{5-r}) \text{ if } x \in \Gamma_r \text{ and } 3 \le r \le 4.$$

The integration by parts shows that

$$B^\dagger(\phi, \psi) = (L\phi, \psi) + \int_\Gamma p(\partial\phi/\partial\nu)\overline{\psi} ds,$$

where $\partial/\partial\nu$ denotes differentiation along the outward normal to Γ, and hence it follows from [**4**, Theorem 2.1, p.11] that

$$B^{\dagger}(u,\psi) = (Lu,\psi) + \int_{\Gamma} p(tr\partial u/\partial\nu)\overline{\psi} ds \quad \text{for} \quad u \in H^2(\Omega).$$

Thus if $u \in V \cap H^2(\Omega)$ and satisfies (2.1b) in the sense of trace, then it is clear that $B(u,\psi) = (Lu,\psi)$ for every $\psi \in C^{\dagger}(\Omega)$, and since $C^{\dagger}(\Omega)$ is dense in V, we conclude from [**120**, Theorem 2.1, p.322] that $u \in D(A)$. The final assertion of the theorem is a direct consequence of Theorem 2.2 since we have shown above that for $u \in D(A)$, u is a solution of (2.8) corresponding to Au.

Theorem 2.5. Let $\lambda^{\dagger} = (\nu,\mu)$ be an eigenvalue of the system (1.1-4). Then μ is an eigenvalue of the problem (2.1) and $\psi^{\star}(x,\lambda^{\dagger})$ a corresponding eigenvector.

Proof. The existence theory for ordinary differential equations assures us that $\phi_r'(x_r,\lambda^{\dagger})$ is absolutely continuous in $0 \le x_r \le 1$ and $\phi''(x_r,\lambda^{\dagger})$ is essentially bounded in $0 < x_r < 1$ for $r = 1,2$, and hence it follows from this fact that $\Psi(x) = \psi^{\star}(x,\lambda^{\dagger}) \in H^2(\Omega)$. Moreover, since $\phi_r(x_r,\lambda^{\dagger})$ satisfies (1.2r) for $r = 1,2$, we also conclude that $\Psi \in V$ and satisfies the boundary condition (2.1b). Thus it follows from Theorem 2.4 that $\Psi \in D(A)$ and $A\Psi = L\Psi$ in $\mathcal{H}$. Finally, the arguments leading to (2.1a) (taking now $\lambda_1 = \nu, \lambda_2 = \mu$, and $y_r = \phi_r(x_r,\lambda^{\dagger})$ for $r = 1,2$) show that $L\Psi = \mu\omega\Psi$ almost everywhere in Ω, and hence $A\Psi = \mu T\Psi$.

Theorem 2.6. Suppose that μ is a real eigenvalue of the problem (2.1) and that $\dim N_{\mu} = n_{\mu} \in \mathbb{N}$. Then there are precisely n_{μ} distinct eigenvalues of the system (1.1-4), say $\{\lambda_j^{\dagger}\}_1^{n_{\mu}}$, whose second components are μ, and moreover, $\{\psi^{\star}(x,\lambda_j^{\dagger})\}_1^{n_{\mu}}$ is a basis of N_{μ}.

Proof. Suppose firstly that $\lambda_j^{\dagger} = (\nu_j,\mu), j = 1,\ldots,p$, are any p distinct eigenvalues of the system (1.1-4) having second component μ (it is important to observe from the Sturm theory that the eigenvalues $\lambda_j^{\dagger}$ are all real - see §1.3). Then it follows from Theorem 2.5 that μ is an eigenvalue of the problem (2.1) and $\psi^{\star}(x,\lambda_j^{\dagger}) \in N_{\mu}$ for $j = 1,\ldots,p$. Let us show that the $\psi^{\star}(x,\lambda_j^{\dagger}), j = 1,\ldots,p$, are linearly independent in $\mathcal{H}$. Accordingly, we may argue as we did in the proof of Theorem 1.2 to show that if $1 \le j,\ k \le p$, and $j \ne k$ then

$$(\nu_j - \nu_k)\int_0^1 A_1(x_1)\phi_1(x_1,\lambda_j^{\dagger})\phi_1(x_1,\lambda_k^{\dagger})dx_1 = 0.$$

Now if the $\psi^*(x, \lambda_j^\dagger)$ are linearly dependent in $\mathcal{H}$, then there exist scalars $\{c_j\}_1^p$, not all zero, such that $\sum_{j=1}^p c_j \psi^*(x, \lambda_j^\dagger) = 0$ for $x \in \Omega$, and hence it follows that

$$c_k \phi_2(x_2, \lambda_k^\dagger) \int_0^1 A_1(x_1)\phi_1^2(x_1, \lambda_k^\dagger)dx_1 = 0 \text{ in } 0 \le x_2 \le 1 \text{ for } k = 1, \ldots, p.$$

Since this implies that $c_k = 0$ for $k = 1, \ldots, p$, we arrive at a contradiction. Thus we have shown that there are at most n_μ distinct eigenvalues of the system (1.1-4) having second component μ and that the corresponding eigenfunctions span a subspace of N_μ of dimension equal to the number of such eigenvalues.

Next let $\{u_j\}_1^{n_\mu}$ be a basis of N_μ. Then for each j, u_j is a solution of the variational problem (2.8) corresponding to $\mu\omega u_j$, and hence it follows from Theorem 2.2 that $u_j \in H^2(\Omega)$ and satisfies the boundary condition (2.1b) in the sense of trace. Moreover, in light of the Sobolev imbedding theorem [**1**, Theorem 5.4, pp.97-98], we also have $\omega u_j \in L^\infty(\Omega)$. Thus it follows from Theorem 2.3 that $u_j \in W^{2,p}(\Omega)$ for any p satisfying $0 < p < \infty$, and so we conclude from the Sobolev imbedding theorem that we may, by modifying u_j on a set of measure zero and extending u_j to all of I^2 by continuity, henceforth suppose that $u_j \in C^1(I^2)$. Note also from the fact that u_j satisfies the boundary condition (2.1b) in the sense of trace that u_j satisfies the boundary conditions (1.2) and (1.4).

Let $\{\Lambda_i\}_0^\infty$ and $\{\eta_i(x_1)\}_0^\infty$ denote the eigenvalues and corresponding eigenfunctions, respectively, of the eigenvalue problem (1.1-2) when λ_2 is held fixed at the value μ, where $-\infty < \Lambda_0 < \Lambda_1 < \ldots, \Lambda_i \to \infty$ as $i \to \infty$, and the η_i are supposed normalized so that $\int_0^1 A_1 \eta_i \eta_k dx_1 = \delta_{ik}$, where δ_{ik} is the Kronecker delta. If we now fix our attention upon a particular u_j and write u for u_j, then it follows from [**55**, Theorem 4.2, p.199 and Problem 11, p.203] that

$$u(., x_2) = \sum_{i=0}^{\infty} \eta_i(.)\chi_i(x_2) \text{ in } L^2(I_1) \tag{2.30}$$

for every $x_2 \in \overline{I_2}$, where $I_r = \{x_r | 0 < x_r < 1\}$ for $r = 1, 2$, and

$$\chi_i(x_2) = \int_0^1 A_1(x_1)\eta_i(x_1)u(x_1, x_2)dx_1.$$

Moreover, if we fix our attention upon a particular i and let $' = d/dx_2$, then it is an immediate consequence of the properties of u cited above that $\chi_i \in C^1(\overline{I_2})$,

$$\chi_i'(x_2) = \int_0^1 A_1(x_1)\eta_i(x_1)D_2u(x_1, x_2)dx_1,$$

and χ_i satisfies the boundary condition (1.4). Also, if $\phi \in C_0^\infty(I_2)$ and if for $n \geq 3$, $\varsigma_n \in C_0^\infty(I_1)$ with $0 \leq \varsigma_n(x) \leq 1$, $\varsigma_n(x_1) = 1$ for $1/n \leq x_1 \leq 1 - 1/n$ and $\varsigma_n(x_1) = 0$ for $x_1 < 1/2n$ and $x_1 > 1 - 1/2n$, then

$$\begin{aligned}\int_0^1 \phi''\chi_i dx_2 &= -\int_0^1 \phi'\chi_i' dx_2 = -\lim_{n\to\infty}\int_0^1 A_1\eta_i\varsigma_n\left(\int_0^1 \phi' D_2 u dx_2\right)dx_1 \\ &= -\lim_{n\to\infty}\int_\Omega D_2(\varsigma_n\phi)\cdot A_1\eta_i D_2 u dx = \lim_{n\to\infty}\int_\Omega \varsigma_n\phi A_1\eta_i D_2^2 u dx \\ &= \lim_{n\to\infty}\int_0^1 A_1\eta_i\varsigma_n\left(\int_0^1 \phi D_2^2 u dx_2\right)dx_1 = \int_0^1 \phi\left(\int_0^1 A_1\eta_i D_2^2 u dx_1\right)dx_2,\end{aligned}$$

where we have used the fact that $u \in H^2(\Omega)$ as well as Fubini's theorem and Lebesgue's dominated convergence theorem. Thus it follows that $\chi_i \in H^2(I_2)$ (where we extend the definition of $H^m(G)$ given in §2.2 to include the 1-dimensional case) and its distributional derivative of the second order is given by

$$\int_0^1 A_1(x_1)\eta_i(x_1)D_2^2 u(x_1, x_2)dx_1.$$

By arguing as in the proof of Theorem 2.7 of [**134**, pp.73-74] we can show that χ_i' is absolutely continuous in $\overline{I_2}$ and that χ_i'' equals the distributional derivative of χ_i' almost everywhere in I_2.

If $L_2^\dagger$ denotes the differential expression on the left-hand side of (1.3) when $\lambda_1 = \Lambda_i$ and $\lambda_2 = \mu$, then let us now show that $L_2^\dagger\chi_i = 0$ almost everywhere in I_2. Accordingly, for $r = 1, 2$, let us introduce in $H^2(\Omega)$ the differential operators

$$L_r(x_r, D_r) = D_r p_r(x_r) D_r + (-1)^{r-1}(\Lambda_i A_r(x_r) - \mu B_r(x_r)) - q_r(x_r).$$

Then

$$A_2 L_1 u + A_1 L_2 u = -Lu + \mu\omega u = 0 \text{ in } \mathcal{H}$$

since we know that in $\mathcal{H}$, $Au = Lu$ (see Theorem 2.4). Hence it follows from this fact and our above results that if $\phi \in C_0^\infty(I_2)$, then

$$\int_{I_2}\phi L_2^\dagger\chi_i dx_2 = \int_\Omega \phi\eta_i A_1 L_2 u dx = -\int_\Omega \phi\eta_i A_2 L_1 u dx. \tag{2.31}$$

On the other hand we know from [**134**, Theorem 2.7, pp.73-74] that for almost every $x_2 \in I_2$, $p_1(x_1)D_1u(x_1,x_2)$ is absolutely continuous in $\overline{I_1}$ and

$$d(p_1(x_1)D_1u(x_1,x_2))/dx_1 = (D_1p_1D_1u)(x_1,x_2) \tag{2.32}$$

almost everywhere in I_1, where the expression on the right-hand side of (2.32) denotes the distributional derivative of p_1D_1u with respect of x_1. Thus if we fix our attention upon the expression in the right-hand side of (2.31), make use of Fubini's theorem, integrate by parts, and appeal to the boundary condition (1.2), then we see that this expression is just

$$-\int_{I_2} \phi A_2 \left(\int_{I_1} \left[(p_1\eta_i')' + (\Lambda_i A_1 - \mu B_1 - q_1)\eta_i \right] u dx_1 \right) dx_2,$$

where $' = d/dx_1$, and hence $\int_{I_2} \phi L_2^\dagger \chi_i dx_2 = 0$. Since ϕ is arbitrary, we conclude that $L_2^\dagger \chi_i = 0$ almost everywhere in I_2, which is what we wanted to prove.

Thus we have shown that if χ_i is not identically zero, then $\eta_i\chi_i$ is an eigenfunction of the system (1.1-4) corresponding to the eigenvalue (Λ_i, μ); and in light of what was said at the beginning of the proof concerning such eigenfunctions, we know that there must be at most n_μ distinct values of i for which χ_i is not identically zero. Hence it follows immediately from (2.30) and the fact that $\dim N_\mu = n_\mu$, that the system (1.1-4) admits precisely n_μ distinct eigenvalues having second component μ and that the corresponding eigenfunctions span N_μ. The proof of the theorem is thus complete.

2.6. Comments

Standard references for the theory of Sobolev spaces, as introduced in §2.2, are Adams [**1**], Maz'ja [**127**], and Mizohata [**134**]. Turning to §2.3, the partial differential equation approach to the eigenvalue problem (1.1-4), that is, via an analysis of the elliptic boundary value problem (2.1), was also considered by Hilbert [**106**], Källström and Sleeman [**116, 117**], and Sleeman [**144**] for the case $\gamma > 0$, by Faierman [**75- 78**] and Faierman and Roach [**89, 92**] for the case $\gamma \geq 0$, and by Faierman [**79-83, 87**] and Faierman and Roach [**90**] for the general case. For a further discussion of the partial differential equation method we refer to Chapter 6 of Volkmer's book [**152**].

Fixing our attention next upon §2.4, the proof of Theorem 2.1 follows closely the arguments used in the proof of Theorem 2.2 of [**73**]. The proof of the assertions of

Theorem 2.2-3 concerning the regularity of u are somewhat lengthy. This of course is due to the fact that Γ contains non-regular points, and hence the usual regularity arguments do not suffice in allowing us to establish the asserted regularity of u. We refer to Grisvard [**101**] and Kondrat'ev and Oleinik [**123**] for an elaboration of the difficulties involved in dealing with elliptic problems in non-smooth domains. The proof given here of Theorem 2.2 was inspired by the work [**115**] of Kadlec. For further generalizations of Theorems 2.2-3 we refer to [**84**].

Finally, the proof of Theorem 2.6 (see §2.5) was strongly influenced by the work of Hilbert [**106**], and the proof of this theorem, as well as the proof of Theorem 2.5, follow closely the arguments of [**92**]. An extension of Theorem 2.6 to the k-parameter case is given in Källström and Sleeman [**116, 117**] and Sleeman [**144**], while for a more abstract version of the theorem see Atkinson [**11**, Theorem 7.9.1.] and Binding and Seddighi [**38**, Theorem 4.3]. For the particular case where $\gamma = 0$, so that $\mu = 0$ is an eigenvalue of the problem (2.1), Binding and Siddighi [**38**] have shown, under certain restrictions on the coefficients of the system (1.1-4), how the n_0 of the theorem can be directly determined from the original system (1.1-4).

CHAPTER 3

The decomposition of $\mathcal{H}$

3.1. Introduction

When $\mu = 0$ is an eigenvalue of the problem (2.1), Theorems 1.2 and 2.6 suggest that we should decompose $\mathcal{H}$ into the direct sum of N_0 and the orthogonal complement of TN_0. Whether such a decomposition is possible will depend very much on the behaviour of the operator T on the subspace N_0 of $\mathcal{H}$; and in this chapter we will deal with this decomposition problem. We shall prove in fact that $\mathcal{H}$ can always be decomposed into a direct sum of M_0 and a certain complementary subspace $\mathcal{H}_0$. This decomposition will be of vital importance in the work to come. Indeed, if we put $\mathcal{H} = \mathcal{H}_0$ if $\mu = 0$ is not an eigenvalue of the problem (2.1), then we shall show in the sequel that there is a compact operator K_0 defined in a subspace of $\mathcal{H}_0$ whose characteristic values are precisely the second components of those eigenvalues of the system (1.1-4) having non-zero second components; and it is by an analysis of the spectral properties of K_0 that we will be able to arrive at our main results.

Since the theory of inner product spaces plays an important role in allowing us to arrive at the above mentioned decomposition of $\mathcal{H}$, we introduce in §3.2 some basic concepts from this theory. These concepts are used in §3.3 to prove a fundamental result (Proposition 3.3) which is of importance not only from the point of view of arriving at the required decomposition of $\mathcal{H}$, but also from the point of view of establishing, for example, bounds for the number of non-real eigenvalues of the system (1.1-4) in terms of the number of negative eigenvalues of the operator A. In §3.4 we show that when $\mu = 0$ is an eigenvalue of the problem (2.1), then $\mathcal{H}$ may be decomposed into the direct sum of M_0 and $\mathcal{H}_0$ and we establish certain other results which are required in the sequel. Finally, in §3.5 we conclude the chapter with some comments.

3.2. Some facts from the theory of inner product spaces

Let X be a vector space over $\mathbf{C}$. Then an inner product on X is a complex-valued function $[\ ,\ ]$ defined for all pairs x, y in X so that the conditions

$$\left[c_1x_1 + c_2x_2, y\right] = c_1\left[x_1, y\right] + c_2\left[x_2, y\right]$$

$$\left[y, x\right] = \overline{\left[x, y\right]}$$

are fulfilled for every c_1, c_2 in $\mathbb{C}$ and x_1, x_2, x, y in X. Note that these conditions imply that

$$\left[y, c_1 x_1 + c_2 x_2\right] = \overline{c_1}\left[y, x_1\right] + \overline{c_2}\left[y, x_2\right].$$

A vector space on which there is defined an inner product $[\ ,\ ]$ is called an inner product space.

If x is an element of the inner product space X, then there are three possibilities: either $[x, x] > 0$, or $[x, x] < 0$, or $[x, x] = 0$. Correspondingly, x is said to be positive, negative, or neutral. If X contains positive as well as negative elements, then it is called an indefinite inner product space. If the inner product space is not indefinite, then it is said to be semi-definite. A semi-definite inner produce space X may either be a positive inner product space ($[x, x] \geq 0$ for every $x \in X$), or a negative inner product space ($[x, x] \leq 0$ for every $x \in X$), or a neutral inner product space ($[x, x] = 0$ for every $x \in X$). If X is a positive (resp. negative) inner product space such that $[x, x] = 0$ implies that $x = 0$, then X is called a positive definite (resp. negative definite) inner product space. The inner product $[\ ,\ ]$ is said to be indefinite, positive semi-definite, negative semi-definite, positive definite, or negative definite on X according to whether X is an indefinite, positive, negative, positive definite, or negative definite inner product space, respectively. Analogous definitions hold for subspaces of X.

Two subsets X_1, X_2 of the inner product space X are said to be orthogonal if $[x_1, x_2] = 0$ for every $x_1 \in X_1$ and $x_2 \in X_2$, while the set $\{x \in X | [x, x_1] = 0 \text{ for every } x_1 \in X_1\}$ is called the orthogonal companion of X_1. If X_1 is a subspace of X and X_1^0 denotes the intersection of X_1 with its orthogonal companion, then the inner product is said to be degenerate on X_1, or X_1 is said to be a degenerate subspace of X, if $X_1^0 \neq 0$. If the subspace X_0 of X is the vector sum of linearly independent, pairwise orthogonal subspaces $X_j, j = 1, \ldots, p \in \mathbb{N}$, of X, then we say that X_0 is the orthogonal direct sum of the subspaces X_j. The inner product space is said to be decomposable if it can be represented as the direct sum of a positive definite, a negative definite, and a neutral subspace which are pairwise orthogonal, and every decomposition of this type is called a fundamental decomposition of X. Finally, if X is a decomposable, non-degenerate inner product space, then every fundamental decomposition of X consists only of a positive definite and a negative definite component.

3.3 Preliminaries

We are now going to present some propositions which will play an important role in allowing us to arrive at a suitable decomposition of $\mathcal{H}$. Our first proposition asserts that for any $\lambda \in \mathbb{C}$, the operator $L - \lambda\omega(x)$ (see §2.3 and Theorem 2.4) has the unique continuation property, and for a proof we refer to [**109**, Theorem 2.4].

Proposition 3.1. Suppose that $u \in D(A)$, $\lambda \in \mathbb{C}$, $Lu - \lambda\omega u = 0$ almost everywhere in Ω and u vanishes almost everywhere in a non-empty open subset of Ω. Then $u = 0$.

We have seen in Chapter 2 that V is a Hilbert space with respect to the inner product $(\, , \,)_{1,\Omega}$. However,

Proposition 3.2. When V, considered only as a vector space, is equipped with the inner product $B(\, , \,)$, then it becomes a positive definite, a positive, or an indefinite inner product space according to whether $\gamma > 0$, $\gamma = 0$, or $\gamma < 0$, respectively.

Proof. It is clear that $B(\, , \,)$ is an inner product on V, and since $B(u,u) \geq \gamma \|u\|^2$ for $u \in V$, it follows that V is positive definite or positive according to whether $\gamma > 0$ or $\gamma = 0$. Note that if $\gamma = 0$, then we have the Schwarz inequality: $|B(u,v)|^2 \leq B(u,u)B(v,v)$ for every $u, v \in V$ (see [**47**, Lemma 2.2, p.5]), and hence an easy argument involving Rellich's theorem [**4**, p.30], (2.2), and Proposition 2.1 shows that for this case there is a $u \neq 0$ in V such that $B(u,u) = 0$. Suppose finally that $\gamma < 0$ and $B(u,u) \leq 0$ for every $u \in V$. Then if we consider V only as a vector space and equip it with the norm $\| \, \|$, it follows from (2.2) and Rellich's theorem that the closed unit ball $\|u\| \leq 1$ in V is totally bounded, and hence we arrive at the contradiction that V is finite dimensional. Thus V must be an indefinite inner product space when $\gamma < 0$.

We shall henceforth write N for $N(A)$ and let $n = \dim N$, so that $n > 0$ if and only if $0 \in \sigma(A)$. Moreover, if $\gamma < 0$, then we shall also let n^- denote the number of negative eigenvalues of A counted according to multiplicity. Now let us fix our attention upon the case $\gamma < 0$. Then we know from above that V is an indefinite inner product space with respect to the inner product $B(\, , \,)$. Hence if we consider V as such a space, then we have

Proposition 3.3. If $\mathcal{L}$ is a negative definite subspace of V, then $\dim \mathcal{L} \leq n^-$, while if $\mathcal{L}$ is a negative subspace of V, then $\dim \mathcal{L} \leq n + n^-$.

Proof. In proving the proposition we shall for brevity suppose that $0 \in \sigma(A)$; the proof for the case $0 \in \rho(A)$ is similar. Accordingly, let $\mathcal{M}^-$ denote the span of those eigenspaces corresponding to the distinct negative eigenvalues of A, and where we note that $\dim \mathcal{M}^- = n^-$. Then we observe that $\mathcal{M}^-$ and N are negative definite and neutral subspaces of V, respectively. Let us now show that V admits a fundamental decomposition having neutral component N, negative definite component $\mathcal{M}^-$, and positive definite τ-closed component $\mathcal{M}^+$, where τ denotes the topology on V induced by the norm $\| \ \|_{1,\Omega}$. To this end let $E(\lambda)$ $\big(E(\lambda+0) = E(\lambda)\big)$ denote the spectral family of A. Then it is clear that $\mathcal{M}^- = E(-0)V$, $N = \big(E(0) - E(-0)\big)V$, and that V is the orthogonal direct sum of the subspaces $\mathcal{M}^-$, N, and $\mathcal{M}^+ = \big(I - E(0)\big)V$. Since it follows from the closed graph theorem that $I - E(0)$ is τ-continuous in V and since we know that $D(A)$ is a core of B, we conclude (see Proposition 2.2) that $\mathcal{M}^+$ is τ-closed and $\big(I - E(0)\big)D(A)$ is τ-dense in $\mathcal{M}^+$. Observing that $B(u,u) \geq 0$ for $u \in \big(I - E(0)\big)D(A)$, it follows from Proposition 2.1 that $\mathcal{M}^+$ is a positive subspace of V. Let us next prove that $\mathcal{M}^+$ is actually positive definite. Indeed, if $u \in \mathcal{M}^+$ and $B(u,u) = 0$, then it follows from the Schwarz inequality that $B(u,v) = 0$ for every $v \in \mathcal{M}^+$, and hence we conclude from the above decomposition of V that $B(u,v) = 0$ for every $v \in V$. Thus it follows from [**120**, Theorem 2.1, p.322] that $u \in N \cap \mathcal{M}^+$, and hence $u = 0$.

Next let P^- denote the projection mapping V onto $\mathcal{M}^-$ along $N + \mathcal{M}^+$. Then if $\mathcal{L}$ is a negative definite subspace of V, if $u \in \mathcal{L}$, and if $P^- u = 0$, we conclude from the above decomposition of V that $u = u_0 + u_+$, where $u_0 \in N$ and $u_+ \in \mathcal{M}_+$. Hence $B(u,u) = 0$ and so $u = 0$. Thus $\mathcal{L}$ is isomorphic to a subspace of $\mathcal{M}^-$ under the mapping P^-, and hence $\dim \mathcal{L} \leq n^-$.

Finally, let P^0 denote the projection mapping V onto $\mathcal{M}^- + N$ along $\mathcal{M}^+$. Then if $\mathcal{L}$ is a negative subspace of V, if $u \in \mathcal{L}$, and if $P^0 u = 0$, we must have $u \in \mathcal{M}^+$, and hence $u = 0$. Thus we conclude that $\mathcal{L}$ is isomorphic to a subspace of $\mathcal{M}^- + N$ under the mapping P^0, and hence $\dim \mathcal{L} \leq n + n^-$.

To conclude this section, let us remark for later use of terminology that the projectors $I - E(0)$ and $E(-0)$ introduced in the proof of the above proposition are called the positive and negative spectral projectors of A, respectively.

3.4 The subspace $\mathcal{H}_0$

As a consequence of our previous results we have

Theorem 3.1. $\mu = 0$ is an eigenvalue of the problem (2.1) if and only if either $\gamma = 0$ or $\gamma < 0$ and $0 \in \sigma(A)$. Moreover, if $\mu = 0$ is an eigenvalue of (2.1), then $N_0 = N$.

Proof. If $\mu = 0$ is an eigenvalue of the problem (2.1), then it is clear that $0 \in \sigma(A)$ and $N_0 \subset N$. Conversely, if $0 \in \sigma(A)$, then $\mu = 0$ is an eigenvalue of (2.1) and $N \subset N_0$. On the other hand, we know from [**120**, p.278] that in order for zero to be in the spectrum of A it is necessary that $\gamma \leq 0$ and sufficient that $\gamma = 0$, and hence all the assertions of the theorem now follow.

If we now consider $\mathcal{H}$ only as a vector space and equip it with the inner product $(\, , \,)_T = (T.,.)$, then we have

Theorem 3.2. If either $\gamma = 0$ or $\gamma < 0$ and $0 \in \sigma(A)$, then there are precisely n distinct eigenvalues of the system (1.1-4) whose second components are zero, and if we denote these eigenvalues by $\lambda(j) = (\lambda_1(j), 0)$ and let $\Psi_j(x) = \psi^\star(x, \lambda(j))$ for $j = 1, \ldots, n$, then $\{\Psi_j\}_1^n$ is a basis of N_0. Moreover, $(\Psi_j, \Psi_k)_T = 0$ for $1 \leq j, \ k \leq n, \ j \neq k$, while $(\Psi_j, \Psi_j)_T \neq 0$ for $j = 1, \ldots, n$ if either ω does not assume both positive and negative values of I^2 or if ω assumes both positive and negative values on I^2 and the inner product $(\, , \,)_T$ is not degenerate on N_0. Finally, if ω assumes both positive and negative values on I^2 and $(\, , \,)_T$ is degenerate on N_0, then there exists the integer m, $1 \leq m \leq n$, such that $(\Psi_j, \Psi_j)_T = 0$ for $j = 1, \ldots, m$ and $(\Psi_j, \Psi_j)_T \neq 0$ for $j = (m+1), \ldots, n$ if $m < n$.

In order to prove the theorem we shall need

Notation. We henceforth let $I_\omega^2 = \{x | x \in I^2, \omega(x) = 0\}$ and let $|I_\omega^2|$ denote the (2-dimensional) Lebesgue measure of I_ω^2.

Proof of Theorem 3.2. The initial assertions of the theorem follow from Theorems 2.6 and 3.1, while the assertion that $(\Psi_j, \Psi_k)_T = 0$ for $j \neq k$ follows from Theorem 1.2. Now suppose that either $\omega \geq 0$ or $\omega \leq 0$ on I^2 and that for some j, $1 \leq j \leq n$, $(\Psi_j, \Psi_j)_T = 0$. Then if $|I_\omega^2| = 0$, we arrive at the contradiction that $\Psi_j(x) = 0$ for $x \in \Omega$. If $|I_\omega^2| > 0$, then we have $L\Psi_j = 0$ in $\mathcal{H}$ and Ψ_j vanishes identically in a non-empty open subset of Ω, and hence in light of Proposition 3.1 we again arrive at the contradiction that $\Psi_j(x) = 0$ for $x \in \Omega$. Finally, suppose that ω assumes both positive and negative values in I^2. Then it is clear from the definitions of the terms involved and from what we have already proved, that the inner product $(\, , \,)_T$ is or is not degenerate on N_0 according to whether

$\prod_{j=1}^{n}(\Psi_j, \Psi_j)_T$ is or is not equal to zero. Hence the proof of the theorem is completed by rearranging and relabelling the terms of the sequence $\{\Psi_j\}_1^n$ if necessary.

Theorem 3.3 Suppose that either $\gamma = 0$ or $\gamma < 0$ and $0 \in \sigma(A)$. Then the vectors $T\Psi_1, \ldots, T\Psi_n$ are linearly independent in $\mathcal{H}$.

Proof. Suppose that the assertion of the theorem is false. Then there exists a vector $\Psi \neq 0$ in $D(A)$ such that $L\Psi = 0$ and Ψ vanishes identically in a non-empty open subset of Ω. Hence, in light of Proposition 3.1, we arrive at a contradiction.

Notation. We henceforth write $\dot{+}$ for the direct sum of subspaces of $\mathcal{H}$, while if X is a subspace of $\mathcal{H}$, then we let $X^{\perp}$ denote its orthogonal complement.

Theorem 3.4. Suppose that $\gamma = 0$ or $\gamma < 0$ and $0 \in \sigma(A)$. Suppose also that either ω does not assume both positive and negative values in I^2 or that ω assumes both positive and negative values in I^2 and the inner product $(\ ,\ )_T$ is not degenerate on N_0. Then $N_0 = M_0$ and $\mathcal{H} = M_0 \dot{+} (TM_0)^{\perp}$.

Proof. If $u \in \mathcal{H}$ and $c_j = (u, \Psi_j)_T / (\Psi_j, \Psi_j)_T$ for $j = 1, \ldots, n$, then $u - \sum_{j=1}^{n} c_j \Psi_j \in (TN_0)^{\perp}$. Moreover, if $u_1 \in N_0$, $u_2 \in (TN_0)^{\perp}$, and $u_1 + u_2 = 0$, then $(u_1, \Psi_j)_T = 0$ for $j = 1, \ldots, n$, and hence $u_1 = u_2 = 0$ since $(\Psi_j, \Psi_j)_T \neq 0$ for $j = 1, \ldots, n$. Thus we have shown that $\mathcal{H} = N_0 \dot{+} (TN_0)^{\perp}$. On the other hand if we assume that $N_0 \neq M_0$, then there is a $u \neq 0$ in N_0 and a $v \in D(A)$ such that $Av = Tu$, and hence we arrive at the contradiction that $(u, \Psi_j)_T = 0$ for $j = 1, \ldots, n$. Thus $N_0 = M_0$, and this completes the proof of the theorem.

Next let $[t] (t \in \mathbb{R})$ denote the integer part of t and U the class of functions f on I^2 of the form $f(x) = \sum_{j=0}^{p} f_{j1}(x_1) f_{j2}(x_2)$, where p is a non-negative integer, $f_{jr} (1 \leq r \leq 2)$ is of class C^1 in $0 \leq x_r \leq 1$ and satisfies the boundary condition (1.2r), f'_{jr} $(' = d/dx_r)$ is absolutely continuous in $0 \leq x_r \leq 1$, while f''_{jr} is essentially bounded in $0 < x_r < 1$.

Theorem 3.5. Suppose that either $\gamma = 0$ or $\gamma < 0$ and $0 \in \sigma(A)$. Suppose also that ω assumes both positive and negative on I^2 and the inner product $(\ ,\ )_T$ is degenerate on N_0. Then the system (1.1-4) uniquely determines the sequence of functions χ_{jk}, $j = 1, \ldots, m$, $k = 1, \ldots, p_j$, on I^2, where $p_j = 1$ for $j = 1, \ldots, m$ if $\gamma = 0$, while $p_1 \geq p_2 \geq \cdots \geq p_m \geq 1$ and $\sum_{j=1}^{m} [p_j/2] \leq n^-$ otherwise, such that $\chi_{jk} \in U$, $\chi_{jk} \in$

$D(A)$, and $A\chi_{jk} = T\chi_{j,k-1}$ if $k > 1$, while $A\chi_{j1} = T\Psi_j$. Moreover, $(\chi_{jk}, \Psi_r)_T = 0$ for $r = 1,\dots,n$, $r \neq j$, $(\chi_{jk}, \Psi_j)_T = 0$ if $k < p_j$, and $(\chi_{j,p_j}, \Psi_j)_T \neq 0$. Finally, the χ_{jk} (resp. the $T\chi_{jk}$) together with the Ψ_r (resp. the $T\Psi_r$), $r = 1,\dots,n$, form a linearly independent set in $\mathcal{H}$, and $M_0 = N_0 \dot{+} span\{\chi_{jk}\}$, $\mathcal{H} = M_0 \dot{+} (TM_0)^{\perp}$.

The proof of the theorem will depend upon the following lemma.

Lemma 3.1. Suppose that the hypotheses of Theorem 3.5 are valid and $1 \leq j \leq m$. Then the system (1.1-4) uniquely determines the sequence of functions on $I^2, \{\chi_{jk}\}_{k=1}^{p_j}$, where $p_j = 1$ if $\gamma = 0$ and $1 \leq p_j \leq 2(n + n^-) - 1$ otherwise, having the following properties. For each $k, \chi_{jk} \in U$, $\chi_{jk} \in D(A)$, and $A\chi_{jk} = T\chi_{j,k-1}$, where $\chi_{j0} = \Psi_j$. Moreover, $(\chi_{jk}, \Psi_r)_T = 0$ for $r = 1,\dots,n$, $r \neq j$, $(\chi_{jk}, \Psi_j)_T = 0$ if $k < p_j$, and $(\chi_{j,p_j}, \Psi_j)_T \neq 0$.

Proof. Let $g_{0r}(x_r) = \phi_r(x_r, \lambda(j))$ for $r = 1, 2$. Then we observe that the condition $(\Psi_j, \Psi_j)_T = 0$ assures us that $e_{01} = e_{02}$, where

$$e_{0r} \int_0^1 A_r g_{0r}^2 dx_r = \int_0^1 B_r g_{0r}^2 dx_r \quad \text{for } r = 1, 2, \tag{3.1}$$

and hence if we put $e_0 = e_{0r}$ and let

$$\omega_r(x_r) = B_r(x_r) - e_0 A_r(x_r) \quad \text{for } r = 1, 2, \tag{3.2}$$

then $\int_0^1 \omega_r g_{0r}^2 dx_r = 0$. Note that the condition $(\Psi_j, \Psi_j)_T = 0$ uniquely determines the constant e_0.

For $r = 1, 2$ let g_{1r} denote the solution of the initial value problem

$$L_r y_r = (-1)^{r-1} h_r(x_r),\ 0 \leq x_r \leq 1,\ y_r(0) = y_r'(0) = 0, \tag{3.3}$$

where $h_r(x_r) = \omega_{0r}(x_r) = \omega_r(x_r) g_{0r}(x_r)$, $' = d/dx_r$, and L_r denotes the differential expression on the left-hand side of (1.2r-1) with $\lambda_1 = \lambda_1(j)$ and $\lambda_2 = 0$. Then for each r, g_{1r} is of class C^1 and g_{1r}' is absolutely continuous in $0 \leq x_r \leq 1$, while g_{1r}'' is essentially bounded in $0 < x_r < 1$. Let us now show that g_{1r} satisfies the boundary condition (1.2r); and clearly we need only show this at $x_r = 1$. Accordingly, let $v_r(x_r)$ denote the solution of $L_r y_r = 0$ in $0 \leq x_r \leq 1$ which satisfies $v_r(0) = -\cos\alpha_r$, $p_r(0)v_r'(0) = \sin\alpha_r$. Then putting

$$F_r(f) = f(1)\cos\beta_r - p_r(1)f'(1)\sin\beta_r,$$

we have

$$F_r(g_{1r}) = (-1)^{r-1}\int_0^1 \Big[g_{0r}(x_r)F_r(v_r) - v_r(x_r)F_r(g_{0r})\Big]\omega_{0r}(x_r)dx_r, \tag{3.4}$$

and hence $F_r(g_{1r}) = 0$ since $\int_0^1 g_{0r}\omega_{0r}dx_r = 0$. Thus g_{1r} satisfies $(1.2r)$, and consequently if we let $f_1(x) = \sum_{r=1}^{2} g_{1r}(x_r)g_{0s}(x_s)$, $s = 3 - r$, then it follows that $f_1 \in U$, while Theorem 2.4 shows that $f_1 \in D(A)$ and $Af_1 = Lf_1$ in $\mathcal{H}$. Moreover, since

$$(Lf)(x) = -A_2(x_2)(L_1 f)(x) - A_1(x_1)(L_2 f)(x) \tag{3.5}$$

almost everywhere in Ω for $f \in U$, it follows immediately that $Af_1 = Tf_0$, where $f_0 = \Psi_j$. We also assert that if $1 \le k \le n$ and $k \ne j$, then $(f_1, \Psi_k)_T = 0$. To prove the assertion let $\phi_{kr}(x_r) = \phi_r(x_r, \lambda(k))$ for $r = 1, 2$. Then we may argue as we did in the proof of Theorem 1.2 to show that if $1 \le r \le 2$, if $L_r Y_r = (-1)^{r-1}F_r$ almost everywhere in $0 \le x_r \le 1$, and if Y_r satisfies the boundary condition $(1.2r)$, we have

$$(\lambda_1(j) - \lambda_1(k))\int_0^1 A_r Y_r \phi_{kr}dx_r = \int_0^1 F_r\phi_{kr}dx_r. \tag{3.6}$$

Hence observing that

$$\omega(x) = -A_2(x_2)\omega_1(x_1) + A_1(x_1)\omega_2(x_2), \tag{3.7}$$

we have on putting $s = 3 - r$,

$$\begin{aligned}(f_1, \Psi_k)_T &= \sum_{r=1}^{2}(-1)^{r-1}\Bigg[\int_0^1 A_r g_{1r}\phi_{kr}dx_r \int_0^1 \omega_s g_{0s}\phi_{ks}dx_s + \\ &\quad + \int_0^1 A_r g_{0r}\phi_{kr}dx_r \int_0^1 \omega_s g_{1s}\phi_{ks}dx_s\Bigg] \\ &= (\lambda_1(j) - \lambda_1(k))^{-1}\sum_{r=1}^{2}(-1)^{r-1}\int_0^1 \omega_r g_{0r}\phi_{kr}dx_r \int_0^1 \omega_s g_{0s}\phi_{ks}dx_s = 0,\end{aligned}$$

which proves the assertion.

Let us next fix our attention upon the case $\gamma = 0$. Then we assert that $(f_1, \Psi_j)_T \neq 0$. Indeed, if this is not the case, then $0 = (f_1, \Psi_j)_T = B(f_1, f_1)$, and since $B \geq 0$ on V, it follows from the Schwarz inequality that $B(f_1, v) = 0$ for every $v \in V$, and hence $f_1 \in N$ [**120**, Theorem 2.1, p.322]. Thus $Tf_0 = 0$, and consequently, in light of Proposition 3.1, we arrive at the contradiction that $f_0 = 0$, which proves the assertion. Putting $\chi_{j1} = f_1$, the proof of the lemma for the case $\gamma = 0$ is complete.

For the remainder of the proof we suppose that $\gamma < 0$. Then there are two cases to consider: (1) $(f_1, \Psi_j)_T \neq 0$ and (2) $(f_1, \Psi_j)_T = 0$. If the first case is valid, then the proof of the lemma is completed by taking $p_j = 1$ and $\chi_{j1} = f_1$. On the other hand, if case (2) is valid, then putting $s = 3 - r$ we have

$$
\begin{aligned}
0 = (f_1, \Psi_j)_T &= \sum_{r=1}^{2}(-1)^{r-1}\left[\int_0^1 A_r g_{1r} g_{0r}\,dx_r \int_0^1 \omega_s g_{0s}^2\,dx_s + \right.\\
&\quad \left. + \int_0^1 A_r g_{0r}^2\,dx_r \int_0^1 \omega_s g_{1s} g_{0s}\,dx_s\right]\\
&= \sum_{r=1}^{2}(-1)^{r-1}\int_0^1 A_r g_{0r}^2\,dx_r \int_0^1 \omega_s g_{1s} g_{0s}\,dx_s,
\end{aligned}
$$

and hence $e_{11} = e_{12}$, where

$$
e_{1r}\int_0^1 A_r g_{0r}^2\,dx_r = \int_0^1 \omega_r g_{1r} g_{0r}\,dx_r \quad \text{for } r = 1, 2. \tag{3.8}
$$

Thus the condition $(f_1, \Psi_j)_T = 0$ uniquely determines the constant $e_1 = e_{1r}$, and if we put

$$
\omega_{1r} = \omega_r g_{1r} - e_1 A_r g_{0r} \quad \text{for } r = 1, 2, \tag{3.9}
$$

then $\int_0^1 \omega_{1r} g_{0r}\,dx_r = 0$.

Let us assume that $(f_1, \Psi_j)_T = 0$. Then we may continue with the above arguments to arrive at the following situation. For $i = 1, \ldots, \ell$ and $r = 1, 2$, there exist the unique numbers e_i and functions g_{ir} such that :

1. g_{ir} is the solution of (3.3) with $h_r = \omega_{i-1,r}$, where

$$
\omega_{ir} = \omega_r g_{ir} - A_r \sum_{k=0}^{i-1} e_{i-k} g_{kr} \quad \text{for } i \geq 1, \tag{3.10}
$$

2.

$$\int_0^1 \omega_{ir} g_{0r} dx_r = 0, \tag{3.11}$$

3. g_{ir} satisfies the boundary condition (1.2r),

4.

$$f_i(x) = \sum_{k=0}^{i} g_{k1}(x_1) g_{i-k,2}(x_2) \tag{3.12}$$

belongs to U, $f_i \in D(A)$ and $Af_i = Tf_{i-1}$, while $(f_i, \Psi_k)_T = 0$ for $k = 1, \ldots, n$.

For $r = 1, 2$ let $g_{\ell+1,r}$ denote the solution of (3.3) with $h_r = \omega_{\ell r}$. Then $g_{\ell+1,r}$ is of class C^1 and $g'_{\ell+1,r}$ ($' = d/dx_r$) is absolutely continuous in $0 \le x_r \le 1$, while $g''_{\ell+1,r}$ is essentially bounded in $0 < x_r < 1$. Moreover, (3.4) (with g_{1r} and ω_{0r} replaced by $g_{\ell+1,r}$ and $\omega_{\ell r}$, respectively) and (3.11) assure us that $g_{\ell+1,r}$ satisfies the boundary condition (1.2r). Hence if we define $f_{\ell+1}$ by means of (3.12), with $i = \ell + 1$, then we conclude that $f_{\ell+1} \in U$, while Theorem 2.4 assures us that $f_{\ell+1} \in D(A)$ and $Af_{\ell+1} = Lf_{\ell+1}$ in $\mathcal{H}$. From (3.7), (3.10), and (3.12) we see that on putting $s = 3 - r$

$$\begin{aligned}
\omega f_l &= \sum_{r=1}^{2} (-1)^{r-1} A_r \omega_s \left[g_{0s} g_{\ell r} + \sum_{k=1}^{\ell} g_{ks} g_{\ell-k,r} \right] \\
&= \sum_{r=1}^{2} (-1)^{r-1} A_r \omega_{0s} g_{\ell r} + \sum_{r=1}^{2} (-1)^{r-1} A_r \sum_{k=1}^{\ell} \left[\omega_{ks} + A_s \sum_{j=0}^{k-1} e_{k-j} g_{js} \right] g_{\ell-k,r} \\
&= \sum_{r=1}^{2} -A_r \sum_{k=0}^{\ell} (-1)^{s-1} \omega_{ks} g_{l-k,r},
\end{aligned}$$

and hence it follows from the definitions of the terms involved and (3.12) that

$$(\omega f_\ell)(x) = \sum_{r=1}^{2} -A_r(x_r)(L_s f_{\ell+1})(x) \text{ almost everywhere in } \Omega.$$

Thus, in light of (3.5), we conclude that $Af_{\ell+1} = Tf_\ell$. If $1 \le k \le n$ and $k \ne j$, then it follows from (3.7) and (3.12) that with $s = 3 - r$,

$$(f_{\ell+1}, \Psi_k)_T = \sum_{r=1}^{2} (-1)^{r-1} \sum_{p=0}^{\ell+1} \int_0^1 A_r g_{pr} \phi_{kr} dx_r \int_0^1 \omega_s g_{\ell+1-p,s} \phi_{ks} dx_s,$$

and hence in light of (3.6) we have

$$(\lambda_1(j)-\lambda_1(k))(f_{\ell+1},\Psi_k)_T=\sum_{r=1}^{2}(-1)^{r-1}\sum_{p=0}^{\ell}\int_0^1 \omega_{pr}\phi_{kr}dx_r\int_0^1 \omega_s g_{\ell-p,s}\phi_{ks}dx_s$$

$$=\sum_{r=1}^{2}(-1)^{r-1}\Big[\sum_{p=0}^{\ell}\int_0^1 \omega_r g_{pr}\phi_{kr}dx_r\int_0^1 \omega_s g_{\ell-p,s}\phi_{ks}dx_s -$$

$$-\sum_{p=1}^{\ell}\sum_{j=0}^{p-1}e_{p-j}\int_0^1 A_r g_{jr}\phi_{kr}dx_r\int_0^1 \omega_s g_{\ell-p,s}\phi_{ks}dx_s\Big] \quad \text{(using (3.10))}$$

$$=-\sum_{r=1}^{2}(-1)^{r-1}\sum_{p=1}^{\ell}\sum_{j=1}^{p}e_j\int_0^1 A_r g_{p-j,r}\phi_{kr}dx_r\int_0^1 \omega_s g_{\ell-p,s}\phi_{ks}dx_s$$

$$=-\sum_{j=1}^{\ell}e_j(f_{\ell-j},\Psi_k)_T \quad \text{(using (3.12) and (3.7))}.$$

Thus $(f_{\ell+1},\Psi_k)_T=0$. Finally, if $(f_{\ell+1},\Psi_j)_T\neq 0$, then the proof of the lemma is completed by taking $p_j=\ell+1$ and $\chi_{jk}=f_k$ for $k=1,\dots,p_j$ (the bound for p_j asserted in the lemma will be established below). On the other hand, if $(f_{\ell+1},\Psi_j)_T=0$, then putting $s=3-r$, it follows from (3.7), (3.10-12) that

$$0=\sum_{r=1}^{2}(-1)^{r-1}\Big[\int_0^1 A_r g_{0r}^2dx_r\int_0^1 \omega_s g_{\ell+1,s}g_{0s}dx_s +$$

$$+\sum_{j=1}^{\ell}\sum_{k=0}^{\ell-j}e_{\ell+1-k-j}\int_0^1 A_r g_{jr}g_{0r}dx_r\int_0^1 A_s g_{ks}g_{0s}dx_s\Big]$$

$$=\sum_{r=1}^{2}(-1)^{r-1}\Big[\int_0^1 A_r g_{0r}^2dx_r\int_0^1 \omega_s g_{\ell+1,s}g_{0s}dx_s +$$

$$+\int_0^1 A_s g_{0s}^2dx_s\int_0^1\Big(A_r\sum_{k=1}^{\ell}e_{\ell+1-k}g_{kr}\Big)g_{0r}dx_r +$$

$$+\sum_{k=1}^{\ell-1}\int_0^1 A_s g_{ks}g_{0s}dx_s\int_0^1\Big(A_r\sum_{j=1}^{\ell-k}e_{\ell+1-k-j}g_{jr}\Big)g_{0r}dx_r\Big]$$

$$=\sum_{r=1}^{2}(-1)^{r-1}\Big[\int_0^1 A_r g_{0r}^2dx_r\int_0^1\Big(\omega_s g_{\ell+1,s}-A_s\sum_{k=1}^{\ell}e_{\ell+1-k}g_{ks}\Big)g_{0s}dx_s -$$

$$-\int_0^1 A_s g_{0s}^2dx_s\int_0^1\Big(\omega_r g_{\ell+1,r}-A_r\sum_{k=1}^{\ell}e_{\ell+1-k}g_{kr}\Big)g_{0r}dx_r +$$

$$+\int_\Omega A_s\omega_r f_{\ell+1}\Psi_j dx\Big] \quad \text{(using the definition of } \omega_{1r}),$$

and hence $e_{\ell+1,1} = e_{\ell+1,2}$, where

$$e_{\ell+1,r}\int_0^1 A_r g_{0r}^2 dx_r = \int_0^1 \left(\omega_r g_{\ell+1,r} - A_r \sum_{k=1}^{\ell} e_{\ell+1-k} g_{kr}\right) g_{0r} dx_r \quad \text{for } r = 1,2. \tag{3.13}$$

Thus the condition $(f_{\ell+1}, \Psi_j)_T = 0$ uniquely determines the constant $e_{\ell+1} = e_{\ell+1,r}$, and if we define $\omega_{\ell+1,r}$ by taking $i = \ell+1$ in (3.10), then (3.11) holds with $i = \ell+1$ for $r = 1,2$. Consequently, if $(f_{\ell+1}, \Psi_j)_T = 0$, then we can continue with the above method to obtain $f_{\ell+2}$.

Finally, we assert that the above method must terminate after a finite number of steps, so that in the previous paragraph only the alternative $(f_{\ell+1}, \Psi_j)_T \neq 0$ can hold and this occurs for $\ell \leq 2(n+n^-)-2$. To see this, let us firstly show that the f_i, $i = 0,\dots,(\ell+1)$ defined above are linearly independent in V. Indeed, if this is not the case, then there is a non-trivial linear combination of these functions, say u_1, which vanishes identically in Ω, and hence $Au_1 = 0$ in $\mathcal{H}$. It follows from the definitions of the terms involved that there is a non-trivial linear combination of the f_i, $i = 0,\dots,\ell$, say u_2, which vanishes identically in a non-empty open subset $\Omega^\#$ of Ω. If u_2 is just a multiple of f_0, then, in light of Proposition 3.1, we arrive at the contradiction that f_0 vanishes identically in Ω. Otherwise, we may use the fact that $Au_2 = 0$ in $L^2(\Omega^\#)$ to show that there is a non-trivial linear combination of the f_i, $i = 0,\dots,(\ell-1)$, say u_3, which vanishes identically in $\Omega^\#$. By arguing with u_3 as we argued with u_2 above, and by repeating this argument if necessary, we finally arrive at the contradiction that f_0 vanishes identically in Ω. Thus we have shown that the f_i, $i = 0,\dots,(\ell+1)$, are linearly independent in V, and moreover, it follows easily from the definitions of the terms involved that $B(f_k, f_r) = 0$ if $k + r \leq \ell + 1$. Hence if p denotes the integer part of $(\ell+1)/2$, then $\text{span}\{f_i\}_0^p$ is a neutral subspace of V with respect to the inner product $B(\ ,\ )$, and so $p + 1 \leq n + n^-$ by Proposition 3.3, which proves the assertion and completes the proof of the lemma.

Proof of Theorem 3.5. For $j = 1,\dots,m$ let χ_{jk} and the p_j be defined according to Lemma 3.1. Then for $\gamma < 0$ we shall henceforth suppose that $p_1 \geq p_2 \geq \cdots \geq p_m \geq 1$ since this can always be achieved if necessary by rearranging the Ψ_j and relabelling them suitably. Now let us firstly prove the theorem for the case $p_1 = 1$. Accordingly, suppose that there exist constants $\{c_j\}_1^n$, $\{c_{j1}\}_1^m$, not all zero, such that

$$\sum_{j=1}^{n} c_j\Psi_j + \sum_{j=1}^{m} c_{j1}\chi_{j1} = 0 \text{ in } \mathcal{H}. \tag{3.14}$$

If we denote the expression on the left-hand side of (3.14) by u and consider the inner product $(u, T\Psi_k)$ for $1 \le k \le m$, then we conclude from Theorem 3.2 and Lemma 3.1 that $c_{k1} = 0$. Similarly we can show that $c_j = 0$ for $j = (m+1), \dots, n$ if $m < n$. Thus we now have $\sum_{j=1}^{m} c_j\Psi_j = 0$ in $\mathcal{H}$, and since the Ψ_j form a linearly independent set $\mathcal{H}$, we arrive at the contradiction that all the c_j and c_{j1} are zero. Hence the Ψ_j together with the χ_{j1} form a linearly independent set in $\mathcal{H}$, and similar arguments together with Proposition 3.1 show that the $T\Psi_j$ together with the $T\chi_{j1}$ also form a linearly independent set in $\mathcal{H}$. Let $M = N_0 \dot{+} span\{\chi_{j1}\}$ and for $f \in \mathcal{H}$ let

$$c_{j1}(f) = (f, \Psi_j)_T / (\chi_{j1}, \Psi_j)_T \text{ for } j = 1, \dots, m.$$

$$c_j(f) = \left[(f, \chi_{j1})_T - \sum_{k=1}^{m} c_{k1}(\chi_{k1}, \chi_{j1})_T\right] / (\Psi_j, \chi_{j1})_T \text{ for } j = 1, \dots, m, \tag{3.15}$$

$$c_j(f) = (f, \Psi_j)_T / (\Psi_j, \Psi_j)_T \text{ for } j = (m+1), \dots, n \text{ if } m < n.$$

Then the above arguments show that f has the unique representation

$$f = \sum_{j=1}^{n} c_j(f)\Psi_j + \sum_{j=1}^{m} c_{j1}(f)\chi_{j1} + f_0,$$

where $f_0 \in (TM)^\perp$, and hence it follows that $\mathcal{H} = M \dot{+} (TM)^\perp$. In light of what has already been proved and since clearly $M \subset M_0$, we see that the proof of the theorem for the case under consideration here will be complete once we have shown that $M_0 \subset M$. Hence suppose that $M_0 \not\subset M$. Then there exists a $u \in M_0 \backslash M$ and a $v \in M$ such that $Au = Tv$. Since $u = u_1 + u_2$, where $u_1 \in M$ and $u_2 \in (TM)^\perp \cap D(A)$, it follows that $Au_2 = Tz$, where $z \in M$, and consequently we conclude from the fact that $Au_2 \in M^\perp$ that $z \in M \cap (TM)^\perp$, and so $z = 0$. Thus u_2, and hence u, belong to M, which is a contradiction.

Suppose next that $p_1 > 1$. Then there exists the integer m_1, $1 \le m_1 \le m$, such that $p_j > 1$ for $1 \le j \le m_1$ and $p_j = 1$ for $m_1 < j \le m$ if $m_1 < m$. If I_1 denotes the set of tuples (j,k), $j = 1, \dots, m_1$, $k = 1, \dots, (p_j - 1)$, then let us show that

$$\det\big((\chi_{jk}, \chi_{rs})_T\big)_{\substack{(j,k)\in I_1 \\ (r,s)\in I_1}} \neq 0. \tag{3.16}$$

Indeed, if this is not the case, then there is a non-trivial linear combination of the $\chi_{jk}, (j,k) \in I_1$, say u_1, such that $(u_1, \chi_{jk})_T = 0$ for $(j,k) \in I_1$ and $(u_1, \Psi_j)_T = 0$ for $j = 1, \ldots, n$. Hence there is a non-trivial linear combination, say u_2, of the χ_{jk}, $j = 1, \ldots, m_1$, $k = 0, \ldots, (p_j - 2)$, where $\chi_{j0} = \Psi_j$, such that $(u_2, \chi_{jk})_T = 0$ for $j = 1, \ldots, m_1$, $k = 1, \ldots, p_j$, and $(u_2, \Psi_j)_T = 0$ for $j = 1, \ldots, n$. If u_2 is a linear combination of only the Ψ_j, $j = 1, \ldots, m_1$, then, in light of Lemma 3.1, we arrive at a contradiction. Otherwise we continue with the above argument to arrive at the contradiction that there is a non-trivial combination of the Ψ_j, $j = 1, \ldots, m_1$, say u_p such that $(u_p, \chi_{j,p_j})_T = 0$ for $j = 1, \ldots, m_1$.

On account of (3.16) we are led to assert that the χ_{jk} together with the Ψ_j form a linearly independent set in $\mathcal{H}$. Indeed, if this is not the case there exist the constants $\{c_j\}_1^n$, $\{c_{jk}\}$, $j = 1, \ldots, m$, $k = 1, \ldots, p_j$, not all zero, such that

$$\sum_{j=1}^{n} c_j \Psi_j + \sum_{j=1}^{m} \sum_{k=1}^{p_j} c_{jk} \chi_{jk} = 0 \text{ in } \mathcal{H}.$$

The arguments used at the beginning of the proof show that $c_j = 0$ for $j = (m + 1), \ldots, n$ if $m < n$ and $c_{j,p_j} = 0$ for $j = 1, \ldots, m$, while (3.16) and Lemma 3.1 show that $c_{jk} = 0$ for $j = 1, \ldots, m_1$, $k = 1, \ldots, (p_j - 1)$. Hence we arrive at the contradiction that $\sum_{j=1}^{m} c_j \Psi_j = 0$ in $\mathcal{H}$, which proves the assertion. Similar arguments together with Proposition 3.1 show that the $T\chi_{jk}$ together with the $T\Psi_j$ also form a linearly independent set in $\mathcal{H}$.

Let $M = N_0 \dot{+} span\{\chi_{jk}\}$ and for $f \in \mathcal{H}$ let:

1. $c_{j,p_j}(f) = (f, \Psi_j)_T / (\chi_{j,p_j}, \Psi_j)_T$ for $j = 1, \ldots, m$,
2. $\{c_{jk}(f)\}_{(j,k)\in I_1}$ denote the unique solution of the simultaneous equations

$$\sum_{(j,k)\in I_1} c_{jk}(f)(\chi_{jk}, \chi_{rs})_T = (f, \chi_{rs})_T - \sum_{j=1}^{m} c_{j,p_j}(f)(\chi_{j,p_j}, \chi_{rs})_T, (r,s) \in I_1, \quad (3.17)$$

3. $c_j(f) = \left[(f, \chi_{j,p_j})_T - \sum_{r=1}^{m} \sum_{s=1}^{p_r} c_{rs}(f)(\chi_{rs}, \chi_{j,p_j})_T\right] / (\Psi_j, \chi_{j,p_j})_T$ for $j = 1, \ldots, m$,
4. $c_j(f) = (f, \Psi)_T / (\Psi_j, \Psi)_T$ for $j = (m+1), \ldots, n$ if $m < n$.

Then the above arguments show that f has the unique representation

$$f = \sum_{j=1}^{n} c_j(f) \Psi_j + \sum_{j=1}^{m} \sum_{k=1}^{p_j} c_{jk}(f) \chi_{jk} + f_0,$$

where $f_0 \in (TM)^{\perp}$, and hence it follows that $\mathcal{H} = M \dot{+} (TM)^{\perp}$. We may now argue as we did before to show that $M_0 = M$, and consequently in light of what has already been proved, we see that the proof of the theorem will be complete once we have obtained the asserted bound for the p_j. Accordingly, let $\chi_{j0} = \Psi_j$ for $j = 1, \ldots, n$ and let $p_j = 0$ for $j = (m+1), \ldots, n$ if $m < n$. Then we already know that $\chi_{jk}, j = 1, \ldots, n,\ k = 0, \ldots, p_j$, form a linearly independent set in V, and moreover, it follows easily from the definitions of the terms involved that $B(\chi_{jk}, \chi_{rs}) = 0$ for $1 \leq j,\ r \leq n,\ 0 \leq k \leq [p_j/2],\ 0 \leq s \leq [p_r/2]$. Hence the $\chi_{jk},\ j = 1, \ldots, n,\ k = 0, \ldots, [p_j/2]$, span a neutral subspace of V with respect to the inner product $B(\ ,\ )$, and so it follows from Proposition 3.3 that $\sum_{j=1}^{m} [p_j/2] + n \leq n + n^-$. Since this gives the required bound, the proof of the theorem is complete.

<u>Remark 3.1.</u> We have defined the eigenvalues, eigenvectors, and associated vectors of the problem (2.1) as those of the pencil $A - \lambda T$ (see [**98**, p.268], [**126**, pp.56-57]). From the point of view of multiparameter spectral theory this is not quite satisfactory since we have not as yet completely described how these terms are related to the system (1.1-4). Theorems 2.5-6 do however furnish us with some information concerning the relationship between the eigenvalues of the system (1.1-4) and those of the pencil $A - \lambda T$, while Theorem 2.6 gives us the result, which is of special importance in multiparameter spectral theory, that if μ is a real eigenvalue of the pencil $A - \lambda T$, then N_μ has a basis consisting of decomposable tensors of $\mathcal{H} = \mathcal{H}^1 \otimes \mathcal{H}^2$, namely the eigenfunctions of the system (1.1-4) (here $\mathcal{H}^r$ denotes the Hilbert space $L^2(0 < x_r < 1)$). The importance of Theorem 3.5 lies in the fact that it shows us that at the eigenvalue $\lambda = 0$ of the pencil $A - \lambda T$, each associated vector is a finite linear combination of decomposable tensors of $\mathcal{H}^1 \otimes \mathcal{H}^2$, while we have shown in the proof of Lemma 3.1 just how the components of each of these latter tensors are related to the system (1.1-4). Further results pertaining to the relationship of the eigenvalues, eigenvectors, and associated vectors of the pencil $A - \lambda T$ to the system (1.1-4) will be given in §§6.5-6.

If either $\gamma = 0$ or $\gamma < 0$ and $0 \in \sigma(A)$, then let $\mathcal{H}_0 = (TM_0)^{\perp}$, $V_0 = V \cap \mathcal{H}_0$, $B_0 = B|V_0$, $A_0 = A|\mathcal{H}_0$, $T_0 = T|\mathcal{H}_0$, and let P_0 denote the orthoprojector mapping $\mathcal{H}$ onto $\mathcal{H}_0$.

<u>Theorem 3.6.</u> Suppose that either $\gamma = 0$ or $\gamma < 0$ and $0 \in \sigma(A)$. Then: (1) $V = M_0 \dot{+} V_0$ and V_0 is a closed subspace of V with respect to the norm $\|\ \|_{1,\Omega}$ and is dense in $\mathcal{H}_0$ with

respect to the norm $\| \ \|$, (2) $D(A) = M_0 \dot{+} D(A_0)$, where $D(A_0) = D(A) \cap \mathcal{H}_0$, and $A_0 : D(A_0) \subset \mathcal{H}_0 \to R(A_0)$ is densely defined in $\mathcal{H}_0$ and closed, with $N(A_0) = 0$, (3) AM_0 and $R(A_0)$ are closed, linearly independent subspaces of $\mathcal{H}$ such that $R(A) = AM_0 \dot{+} R(A_0)$, (4) $\mathcal{H} = TM_0 \dot{+} R(A_0)$ and $R(A_0) = M_0^{\perp}$, (5) the mapping $A_0^{-1} : R(A_0) \to \mathcal{H}_0$ is compact, (6) $R(T_0) \subset R(A_0)$, (7) $N(T) = N(T_0)$, and lastly (8) P_0 maps $R(A_0)$ isomorphically onto $\mathcal{H}_0$.

Proof. Let P denote the projection mapping $\mathcal{H}$ onto $\mathcal{H}_0$ along M_0. Then fixing our attention firstly upon assertion (1), we have $V = (I-P)V \dot{+} PV$, and hence $(I-P)V \subset M_0$ and $PV \subset V_0$. Since $M_0 = (I-P)M_0 \subset (I-P)V$ and $V_0 = PV_0 \subset PV$, we conclude that $(I-P)V = M_0$ and $PV = V_0$. To show that V_0 is a closed subspace of the Hilbert space V, we observe that if $u_n \to u$ in V, where $u_n \in V_0$ for each n, then $u_n \to u$ in $\mathcal{H}$, and hence $u \in V_0$ since $\mathcal{H}_0$ is a closed subspace of $\mathcal{H}$. To show that V_0 is a dense subspace of the Hilbert space $\mathcal{H}_0$, we observe that if $u \in \mathcal{H}_0$ and ϵ is any positive number, then there is a $v \in V$ such that $\|u - v\| < \epsilon$, and hence $\|u - Pv\| \leq \epsilon \|P\|_{\mathcal{H}}$, where $\| \ \|_{\mathcal{H}}$ denotes the norm in $\mathcal{L}(\mathcal{H})$. Turning to assertion (2), we can show as before that $D(A) = (I-P)D(A) \dot{+} PD(A)$, $(I-P)D(A) = M_0$ and $PD(A) = D(A_0)$, while it follows from the definition of A_0 that $N(A_0) = 0$. Since $D(A)$ is a dense subspace of $\mathcal{H}$, we can show as before that $D(A_0)$ is a dense subspace of $\mathcal{H}_0$. If $\{u_n\}$ is a sequence in $D(A_0)$ such that $u_n \to u$ in $\mathcal{H}_0$ and $A_0 u_n \to v$ in $R(A_0)$, then $u \in D(A) \cap \mathcal{H}_0$ and $Au = v$ since A is closed, and hence we conclude that A_0 is closed.

Turning to assertion (3), we see that the assertion is certainly true if $M_0 = N_0$, and hence let us fix our attention upon the case $M_0 \neq N_0$ and show firstly that AM_0 and $R(A_0)$ are linearly independent subspaces of $\mathcal{H}$. Indeed, if this is not the case, then there is a $u \neq 0$ in M_0 and a $v \in D(A_0)$ such that $Tu + A_0 v = 0$. On the other hand, it is easy to see from Theorem 3.5 that $A_0 v \in M_0^{\perp}$, and hence we arrive at the contradiction that $u \in \mathcal{H}_0$. Thus AM_0 and $R(A_0)$ are linearly independent subspaces of $\mathcal{H}$, and since $R(A)$ and AM_0 are closed subspaces of $\mathcal{H}$ and $R(A) = AM_0 \dot{+} R(A_0)$, we conclude from [**100**, Theorem IV.1.12, p.100] that $R(A_0)$ is also closed in $\mathcal{H}$, which completes the proof of assertion (3).

Fixing our attention next upon assertion (4), let us firstly show that $\mathcal{H} = TM_0 \dot{+} M_0^{\perp}$. Accordingly, if we assume that $TM_0 + M_0^{\perp}$, which is a closed subspace of $\mathcal{H}$, is actually a proper subspace, then we arrive at the contradiction that $M_0 \cap (TM_0)^{\perp} \neq 0$. Furthermore, if we suppose that $TM_0 \cap M_0^{\perp} \neq 0$, then we arrive at the contradiction that there is a $u \neq 0$ in $\mathcal{H}$ such that $(u, v) = 0$ for every $v \in M_0$ and for every $v \in (TM_0)^{\perp}$. Thus the required decomposition of $\mathcal{H}$ is established, and hence if $M_0 = N_0$, then the proof of the assertion is complete. Let us now suppose that $M_0 \neq N_0$ and show that $M_0^\perp = R(A_0)$. Then if we observe that $M_0^\perp \subset R(A)$ and $AM_0 \subset TM_0$, it follows from assertion (3) that $M_0^\perp \subset TM_0 + R(A_0)$. On the other hand we know from above that $R(A_0) \subset M_0^\perp$, and consequently we must have $R(A_0) = M_0^\perp$.

To prove assertion (5) let $\{v_n\}$ be a bounded sequence in $R(A_0)$, let $\lambda \in \rho(A)$, and let $u_n = (A - \lambda I)^{-1} v_n$. Since $(A - \lambda I)^{-1}$ is compact, $\{u_n\}$ contains a convergent subsequence, which by relabelling suitably we may again denote by $\{u_n\}$. Moreover, if $P^\dagger$ denotes the projection mapping $\mathcal{H}$ onto $R(A_0)$ along TM_0, then

$$A(I-P)u_n + A_0 P u_n = v_n + \lambda(I - P^\dagger)u_n + \lambda P^\dagger u_n,$$

and since $A(I-P)u_n$ and $(I-P^\dagger)u_n$ are contained in TM_0, we conclude from assertion (4) that $Pu_n - \lambda A_0^{-1} P^\dagger u_n = A_0^{-1} v_n$. On the other hand, assertion (2) and the closed graph theorem assure us that A_0^{-1} is bounded, and hence $\{Pu_n - \lambda A_0^{-1} P^\dagger u_n\}$ is a convergent sequence in $\mathcal{H}_0$, which proves the assertion.

Assertion (6) follows from assertion (4) and the fact that $u \in (TM_0)^\perp$ for $u \in \mathcal{H}_0$, and hence $Tu \in M_0^\perp$. To prove assertion (7), it is clear that we need only consider the case $N(T) \neq 0$ and show that $N(T) \subset N(T_0)$. Hence if $u \in N(T)$ and $u = u_1 + u_2$, where $u_1 \in M_0$ and $u_2 \in \mathcal{H}_0$, then it follows from assertions (4) and (6) that $Tu_i = 0$ for $i = 1,2$. On the other hand we know from Theorems 3.3 and 3.5 that the mapping $T : M_0 \to TM_0$ is an injection, and hence $u_1 = 0$ and $u = u_2$. Finally, we see from assertion (4) that the mapping $P_0 : R(A_0) \to \mathcal{H}_0$ is bijective, and hence assertion (8) follows immediately from the closed graph theorem.

If either $\gamma > 0$ or $\gamma < 0$ and $0 \in \rho(A)$, then we shall henceforth write $\mathcal{H}_0$ for $\mathcal{H}$, V_0 for V, B_0 for B, A_0 for A, T_0 for T, and let P_0 denote the identity operator in $\mathcal{H}$.

<u>Theorem 3.7.</u> $B_0(u,v)$ is a closed, densely defined, bounded from below symmetric form in $\mathcal{H}_0$ which is coercive over V_0 and satisfies $|B_0(u,v)| \leq c\|u\|_{1,\Omega}\|v\|_{1,\Omega}$ for $u,v \in V_0$, where c denotes a positive constant, while $A_0^\dagger = P_0 A_0$ is precisely the selfadjoint operator in $\mathcal{H}_0$ that is associated with the form B_0 and $0 \in \rho(A_0^\dagger)$. Moreover, if $\gamma \geq 0$, then B_0 is positive definite on V_0, while if $\gamma < 0$ and $0 \in \rho(A)$ or if $0 \in \sigma(A)$ and either ω does not assume both positive and negative values on I_2 or ω assumes both positive and negative values on I^2 and the inner product $(\ ,\ )_T$ is not degenerate on N_0, then B_0 is indefinite on V_0. If $\gamma < 0$, $0 \in \sigma(A)$, ω assumes both positive and negative values on I^2, and the inner product $(\ ,\ )_T$ is degenerate on N_0, then B_0 is either positive definite or indefinite

on V_0. Finally, if B_0 is positive definite on V_0, then $B_0(u,u) \geq k\|u\|^2_{1,\Omega}$ for $u \in V_0$, where k denotes a positive constant.

Proof. The initial assertions concerning B_0 follow from Propositions **2.1-2** and Theorem 3.6. To prove the assertions concerning $A_0^\dagger$, it is clear that we need only consider the case where $0 \in \sigma(A)$. Accordingly, it follows from Theorem 3.6 and the closed graph theorem that $A_0^\dagger$ is a selfadjoint operator in $\mathcal{H}$ and $0 \in \rho(A_0^\dagger)$. On the other hand, if $u \in D(A_0^\dagger) = D(A_0)$, then $(A_0^\dagger u, v) = B_0(u,v)$ for every $v \in V_0$, and hence it follows from [**120**, Theorem 2.1, p.322] that $A_0^\dagger \subset A^\dagger$, where $A^\dagger$ denotes the selfadjoint operator in $\mathcal{H}_0$ associated with B_0. Thus $A_0^\dagger = A^\dagger$.

If $\gamma > 0$, then we know from Proposition 3.2 that B_0 is positive definite on V_0, while (2.2) shows us that $B_0(u,u) \geq c_0(1 + c_1\gamma^{-1})^{-1}\|u\|^2_{1,\Omega}$ for $u \in V_0$. If $\gamma = 0$, then we know from Proposition 3.2 that B_0 is positive semi-definite on V_0. Let us now show that B_0 is actually positive definite on V_0. Indeed, if $u \in V_0$ and $B_0(u,u) = 0$, then it follows from the Schwarz inequality that $B_0(u,v) = 0$, for every $v \in V$. Thus $u \in N(A_0^\dagger)$, and hence $u = 0$. If γ_0 denotes the lower bound of B_0, then we have just seen that $\gamma_0 \geq 0$; and we assert that actually $\gamma_0 > 0$. Indeed, if this is not the case, then, in light of [**120**, p.278 and Theorem 2.6, p.323], we arrive at the contradiction that $0 \in \sigma(A_0^\dagger)$. As a consequence of (2.2), it now follows that

$$B_0(u,u) \geq c_0(1 + c_1\gamma_0^{-1})^{-1}\|u\|^2_{1,\Omega} \quad \text{for} \quad u \in V_0. \tag{3.18}$$

Turning next to the case $\gamma < 0$, $0 \in \rho(A)$, we see from Proposition 3.2 that for this case, B_0 is indefinite on V_0. We assert that the same result is also true if $\gamma < 0$, $0 \in \sigma(A)$, and either ω does not assume both positive and negative values on I^2 or ω assumes both positive and negative values on I^2 and the inner product $(\ ,\ )_T$ is not degenerate in N_0. Indeed, the assertion follows immediately from the fact that if $v \in V$ and $v = v_1 + v_2$, where $v_1 \in M_0$ and $v_2 \in V_0$, then $B(v,v) = B_0(v_2,v_2)$. Finally, suppose that $\gamma < 0$, $0 \in \sigma(A)$, ω assumes both positive and negative values on I^2, and the inner product $(\ ,\ )_T$ is degenerate on N_0. Then we may argue as we did in the proof of Proposition 3.2 to show that for this case B_0 cannot be negative definite or negative semi-definite on V_0. We also observe that B_0 cannot be positive semi-definite on V_0 since otherwise there is a $u \neq 0$ in V_0 such that $B_0(u,v) = 0$ for every $v \in V_0$, and hence $u \in N(A_0^\dagger)$. Since $N(A_0^\dagger) = 0$, we arrive at a contradiction. Thus B_0 is either indefinite or positive definite on V_0. If B_0 is positive definite on V_0 and γ_0 denotes the lower bound of B_0, then we may argue as we did with the case $\gamma = 0$ to show that $\gamma_0 > 0$ and that (3.18) is valid.

Theorem 3.8. If λ is a non-zero eigenvalue of the problem (2.1), then $M_\lambda \subset D(A_0)$.

Proof. It is clear that we need only prove the theorem for the case $0 \in \sigma(A)$; and in light of Theorems 3.4-6 we see that in order to prove the theorem we have to show $(u,v)_T = 0$ for $u \in M_\lambda$ and $v \in M_0$. Now observe that if $u \in N_\lambda$ and $1 \le j \le n$, then $(u, \Psi_j)_T = \lambda^{-1}(u, A\Psi_j) = 0$, where we refer to Theorem 3.2 for terminology. Moreover, if $\lambda = 0$ is not a semi-simple eigenvalue of the problem (2.1) and the χ_{jk} are the vectors defined in Theorem 3.5, then the equation $(u, \chi_{jk})_T = \lambda^{-1}(u, \chi_{j,k-1})_T$, where $\chi_{j,0} = \Psi_j$, assures us that $(u, \chi_{j,k})_T = 0$ for $j = 1, \ldots, p_j$. Thus we have shown that $(u,v)_T = 0$ for $v \in M_0$.

Suppose next that λ is not a semi-simple eigenvalue of the problem (2.1) and $u \in M_\lambda$, $u \notin N_\lambda$. Then there exist the vectors $\{u_r\}_0^{p-1}$ in M_λ, where $p > 1$, $u_{p-1} = u$, and $u_0 \neq 0$, such that $(A - \lambda T)u_r = Tu_{r-1}$, where $u_{-1} = 0$. Hence if $1 \le j \le n$, then it follows from the equation $(u_r, \Psi_j)_T = -\lambda^{-1}(u_{r-1}, \Psi_j)_T$, that $(u_r, \Psi_j)_T = 0$ for $r = 0, \ldots, (p-1)$. Furthermore, if $\lambda = 0$ is not a semi-simple eigenvalue of (2.1), then an inductive argument involving the foregoing results and the equation $(u_r, \chi_{jk})_T = \lambda^{-1}(u_r, \chi_{j,k-1})_T - \lambda^{-1}(u_{r-1}, \chi_{jk})_T$ assures us that $(u_r, \chi_{jk})_T = 0$ for $r = 0, \ldots, (p-1)$ and $k = 1, \ldots, p_j$. Thus we see that $(u,v)_T = 0$ for $v \in M_0$, and this completes the proof of the theorem.

Finally, for later use we require

Theorem 3.9. Suppose that ω assumes both positive and negative values on I^2. Then when $\mathcal{H}_0$, considered only as a vector space, is equipped with the inner product $(\ ,\ )_T$, it becomes an indefinite inner product space.

Proof. It is clear that we need only prove the theorem for the case $0 \in \sigma(A)$. Accordingly, let us consider $\mathcal{H}$ only as a vector space and equip it with the inner product $(\ ,\ )_T$. Then $\mathcal{H}$ becomes an indefinite inner product space, and as such, Theorem 3.4-5 show that $\mathcal{H}$ is the orthogonal direct sum of the subspace $\mathcal{H}_0$ and the non-degenerate subspace M_0. Moreover, it follows from [**47**, Corollary 11.8, p.26] that M_0 is decomposable, and hence M_0 is the orthogonal direct sum of a positive definite subspace M_0^+ and a negative definite subspace M_0^-. Also from the facts that $(u,v)_T = (P_0T_0u, v)$ for $u, v \in \mathcal{H}_0$ and P_0T_0 is a bounded selfadjoint operator in $\mathcal{H}_0$, it follows from the spectral theorem (see [**47**, Theorem 5.2, p.89]) that $\mathcal{H}_0$ is the orthogonal direct sum of a τ-closed positive definite subspace $\mathcal{H}_0^+$, a τ-closed negative definite subspace $\mathcal{H}_0^-$, and a τ-closed neutral subspace of $\mathcal{H}_0^0$, where τ denotes the topology on $\mathcal{H}_0$ induced by the norm $\| \ \|$. Hence $\mathcal{H}$ admits the fundamental decomposition

$$\mathcal{H} = \mathcal{H}_0^0(\dot{+})(\mathcal{H}_0^+(\dot{+})M_0^+)(\dot{+})(\mathcal{H}_0^-(\dot{+})M_0^-),$$

where $(\dot{+})$ denotes the orthogonal direct sum of subspaces of the indefinite inner product space $\mathcal{H}$.

Let $\mathcal{L}$ be a positive definite subspace of the indefinite inner product space $\mathcal{H}$. Then we may argue as we did in the proof of Proposition 3.3 to show that $\mathcal{L}$ is isomorphic to a subspace of $M_0^+\dot{+}\mathcal{H}_0^+$. Hence, since $\mathcal{H}$ contains positive definite subspaces of arbitrarily large dimensions and since $\dim M_0^+ \leq 2(n+n^-)$ (see Theorems 3.4-5), we conclude that $\mathcal{H}_0^+ \neq 0$. Similar arguments show that $\mathcal{H}_0^- \neq 0$, and this completes the proof of the theorem.

Corollary 3.1. Suppose that ω assumes both positive and negative values in I^2. Then when V_0, considered only as a vector space, is equipped with the inner product $(\ , \ )_T$, it becomes an indefinite inner product space.

Proof. Suppose that the assertion is false. Then V_0 is a semi-definite inner product space with respect to the inner product $(\ , \ )_T$, and hence so is $\mathcal{H}_0$, in light of Theorem 3.6. Since this contradicts Theorem 3.9, the corollary is proved.

3.5. Comments

The theory of inner product spaces discussed in §3.2 will play a fundamental role in our work and standard references for this theory are Azizov and Iokhvidov [**14**], Bognár [**47**], and Iohvidov, Krein, and Langer [**111**].

Turning to §3.3, we began this section by showing that L has the unique continuation property and made essential use of this fact in the following section in arriving at the required decomposition of $\mathcal{H}$. It appears that the unique continuation hypothesis is indispensable in dealing with general elliptic boundary value problems involving an indefinite weight function and we refer to [**85**] for a discussion of this point. Propositions 3.2-3 and their proofs are taken from [**82**, §§4-5], [**83**, §6], and [**90**, Remark 7.1]. Note that in Proposition 3.3 we used some concepts from the theory of inner product spaces to establish bounds for $\dim \mathcal{L}$. It is important to observe that the proposition asserts that if $\mathcal{L}$ is a negative definite (resp. negative) subspace of V (considered as an indefinite inner product space), then $\dim \mathcal{L}$ does not exceed the number of negative eigenvalues (resp. non-positive) eigenvalues, counted according to their multiplicities, of the unweighted elliptic boundary value problem : $Lu = \mu u$ in Ω together with the boundary condition (2.1b). Thus here (and this will also be the case in the sequel) we are able to establish bounds depending upon the number of non-positive eigenvalues of an associated unweighted elliptic boundary value problem. Finally, we note that a further refinement of this last result has been made by Binding and Seddighi [**38**] wherein a bound for the number of negative eigenvalues concerned, n^-, is obtained which depends explicitly upon the original system (1.1-4), while a similar bound for n is also given under certain other restrictions.

Particular cases of the decomposition problem dealt with in §3.4 are considered in Faierman [**77,78,80,82,83**] and Faierman and Roach [**89,90**]. The proofs of Lemma 3.1 and Theorem 3.7 follow closely the arguments of [**87**] and [**81**], respectively, while the proof of Theorem 3.6 is based upon ideas sketched out in [**81,83,90**]. We might also mention that our methods for dealing with the decomposition problem are essentially those used in [**85**, §2] wherein this same problem for general elliptic boundary value problems involving indefinite weights is dealt with. Fixing our attention upon Remark 3.1, the structure of the principal subspaces of the pencil $A - \lambda T$ which arises from a more general multiparameter eigenvalue problem than that considered here has been the subject of investigation in Binding and Seddighi [**38**] and Binding [**24**]. These authors introduce an assumption equivalent to condition (ii) of Assumption 1.1 plus a second assumption, which in [**24**] is equivalent to $|I_\omega^2| = 0$, and which in [**38**] ensures, in our terminology, that $N(T) \cap [R(\tilde{A})]^\perp = 0$, where $\tilde{A} = A|(D(A) \cap N(T))$ (note that in [**38**] more assumptions than those just stated are introduced when dealing with the structure of the principal subspace of the pencil corresponding to $\lambda = 0$ when $\lambda = 0$ is an eigenvalue). This second assumption enables the authors, by means of a shift in the spectral parameter, to arrive at the situation where $0 \in \rho(A)$, and hence guarantee that the set of regular values of the pencil is not empty (see [**126**, p.56] for terminology). Then under the aforementioned hypotheses, these two papers together completely describe the structure of the principal subspaces of the pencil $A - \lambda T$ corresponding to its real eigenvalues. Indeed, it is shown that at the real eigenvalues of the pencil, the principal subspaces have the structure exhibited in Theorems 3.4-5. Note that the results of these works do not include ours since we have not introduced any further assumptions apart from Assumption 1.1. For further discussions concerning the principal subspaces associated with general multiparameter eigenvalue problems we refer to Atkinson [**10**] and [**11**, Chapter 6], Gadzhiev [**97**], and Isaev [**112**]. Finally, for an investigation of the decomposition problem associated with linear pencils in general we refer to Binding [**25**].

CHAPTER 4

The right semi-definite case

4.1. Introduction

We say that the eigenvalue problems (1.1-4) and (2.1) are right semi-definite if either $\omega \geq 0$ or $\omega \leq 0$ in I^2. In this chapter we shall fix our attention upon the right semi-definite case, and to be more precise, we shall suppose throughout that $\omega \geq 0$ in I^2. Note that this assumption involves no loss of generality since the case $\omega \leq 0$ in I^2 can always be reduced to the case under consideration here by means of the transformation $\lambda_1' = \lambda_1$, $\lambda_2' = -\lambda_2$ in the parameters λ_1, λ_2 of (1.1) and (1.3). Then as a consequence of the results of Chapters 1-3, we are now in a position to establish some important information concerning the spectral properties of the problem (2.1) and the system (1.1-4). Accordingly, in §4.2 below we prove some results concerning the eigenvalues of the problem (2.1) and show that the eigenvectors are complete in certain function spaces. In §4.3 we make use of theorems 2.5-6 to obtain analogous results for the eigenvalues and eigenfunctions of the system (1.1-4) and we also deal with the uniform convergence of the eigenfunction expansion associated with (1.1-4). Finally, in §4.4 we conclude the chapter with some comments.

4.2. The problem (2.1)

We are now going to present our main results for the problem (2.1). Accordingly, recalling the definitions of §§2.3 and 3.3, we have

Theorem 4.1. The eigenvalues of the problem (2.1) form an denumerably infinite subset of $\mathbb{R}$ having no finite points of accumulation and each eigenvalue is semi-simple and of finite multiplicity. Moreover, the eigenvalues are all positive if $\gamma > 0$, all non-negative if $\gamma = 0$, while if $\gamma < 0$, then there may appear negative eigenvalues, but the total number of such eigenvalues, counted according to multiplicity, does not exceed n^-.

In order to prove the theorem, we require

Definition 4.1. Let X be a complex Hilbert space and H a linear operator in X. Then a complex number μ is called a characteristic value of H if there exists a vector $u \neq 0$ in $D(H)$ such that $\mu H u = u$; u is called a characteristic vector of H corresponding to μ.

Proof of Theorem 4.1. Fixing our attention again upon Theorem 3.6, let us put $P = P_0|R(A_0)$ and $K = A_0^{-1}T_0$. Then it follows that K is a compact operator in $\mathcal{H}_0$, and we assert, moreover, that K is strongly symmetrisable with respect to the positive operator PT_0, i.e., PT_0K is a selfadjoint operator in $\mathcal{H}_0$ and $N(PT_0) \subset N(K)$ (see [**154**, p.371]). To see this, let $u,\ v \in \mathcal{H}_0$. Then it follows from Theorems 3.6-7 that $(PT_0Ku, v) = (Ku, T_0v) = ((A_0^\dagger)^{-1}PT_0u, PT_0v) = \left(PT_0u, (A_0^\dagger)^{-1}PT_0v\right) = (u, PT_0Kv)$. Hence, since $N(PT_0) = N(T_0) = N(K)$, the assertion now follows. We conclude from [**154**, p.372] that the eigenvalues of K are all real and that each non-zero eigenvalue is semi-simple. Moreover, if we consider $\mathcal{H}_0$ only as a vector space and equip it with the inner product $(\ ,\ )_T$, then in this inner product space we have: (1) the eigenspace corresponding to a non-zero eigenvalue of K is positive definite and (2) the eigenspaces corresponding to the distinct eigenvalues of K are pairwise orthogonal.

If λ is an eigenvalue of K, then let us denote by $\mathcal{G}_\lambda$ the eigenspace of K corresponding to λ. We are now going to show that the characteristic values of K are precisely the non-zero eigenvalues of the problem (2.1) and that for each characteristic value λ we have $\mathcal{G}_{1/\lambda} = N_\lambda$. Accordingly, let us firstly suppose that $\lambda \neq 0$ is an eigenvalue of the problem (2.1). Then we know from Theorem 3.8 that $N_\lambda \subset D(A_0)$, and hence if $0 \neq u \in N_\lambda$, then $A_0u = \lambda T_0u$. Thus $u = \lambda Ku$, and so we conclude that λ is a characteristic value of K and $N_\lambda \subset \mathcal{G}_{1/\lambda}$. Conversely, if λ is a characteristic value of K and $0 \neq u \in \mathcal{G}_{1/\lambda}$, then $\lambda Ku = u$, and hence we see that $u \in D(A_0)$ and $A_0u = \lambda T_0u$. Thus we conclude that λ is a non-zero eigenvalue of the problem (2.1) and $\mathcal{G}_{1/\lambda} \subset N_\lambda$.

Let us next show that if λ is a non-zero eigenvalue of the problem (2.1) then $M_\lambda = N_\lambda$. Indeed, if this is not the case, then there exist vectors $u \neq 0$ in N_λ and $v \in M_\lambda$ such that $(A - \lambda T)v = Tu$. Moreover, in light of Theorem 3.8 we know that $u, v \in D(A_0)$. Hence $(A_0 - \lambda T_0)v = T_0u$, and so we conclude that $(I - \lambda K)v = Ku$ and $(I - \lambda K)Ku = 0$. Since $1/\lambda$ is a semi-simple eigenvalue of K, this implies that $Ku = 0$, and hence $T_0u = 0$. Thus, since N_λ is positive definite with respect to the inner product $(\ ,\ )_T$, we arrive at the contradiction that $u = 0$.

Now suppose that λ is a non-zero eigenvalue of the problem (2.1) and u a corresponding eigenvector. Then we know from above that $u \in D(A_0)$, and hence it follows that

$$B_0(u, u) = \lambda(u, u)_T. \tag{4.1}$$

Thus, since we have shown that the eigenspaces of the problem (2.1) corresponding to its non-zero eigenvalues are positive definite with respect to the inner product $(\ ,\ )_T$, we conclude from Theorem 3.7 and (4.1) that $\lambda > 0$ if $\gamma \geq 0$, while if $\gamma < 0$, then either $\lambda > 0$ or $\lambda < 0$. Moreover, if there are negative eigenvalues and if $\mathcal{G}_-$ denotes the span of the N_λ for which $\lambda < 0$, then it follows from (4.1) and the fact that the eigenspaces of the problem (2.1) corresponding to its non-zero eigenvalues are pairwise orthogonal with respect to the inner product $(\ , \ )_T$, that $\mathcal{G}_-$ is a negative definite subspace of V with respect to the inner product $B(\ , \ )$. Hence we conclude from Proposition 3.3 that $\dim \mathcal{G}_- \leq n^-$. Furthermore, since it is a simple matter to verify that the eigenspaces of the problem (2.1) are linearly independent in $\mathcal{H}$ (see Theorem 3.4), we have also established that $\Sigma \dim N_\lambda \leq n^-$, where the summation is over those eigenvalues λ of the problem (2.1) for which $\lambda < 0$.

In light of the foregoing results and Theorems 1.1, 2.5-6, 3.1-2, and 3.4, the proof of the theorem is now complete.

Remark 4.1. It was shown in the above proof that the eigenspaces of the problem (2.1) corresponding to its non-zero eigenvalues are positive definite with respect to the inner product $(\ , \ )_T$. By appealing to Theorem 3.2, it is a simple matter to verify that this result remains valid for N_0 if $\lambda = 0$ is an eigenvalue of the problem (2.1) (see Theorem 3.1).

Notation. We henceforth let $\Omega^\# = \Omega \backslash I_\omega^2$, where we refer to the notation preceding the proof of Theorem 3.2 for terminology.

Theorem 4.2. The eigenvectors of the problem (2.1) are complete in $L^2(\Omega^\#)$ and in $L^2(\Omega^\#; \omega(x)dx)$.

Remark 4.2. Before beginning the proof of the theorem let us note that the eigenvectors and associated vectors of the problem (2.1) are members of the function space $\mathcal{H}$, and hence what the theorem asserts is that the restrictions of the eigenvectors to the set $\Omega^\#$, when $\Omega^\# \neq \Omega$, have the properties stated.

Proof of Theorem 4.2. We see from the proof of Theorem 4.1 that the characteristic values and characteristic vectors of K may be arranged into sequences $\{\mu_j\}_1^\infty$ and $\{u_j\}_1^\infty$, respectively, where each characteristic value is repeated according to its multiplicity (i.e., according to the multiplicity of its reciprocal as an eigenvalue of K), $0 < |\mu_1| \leq |\mu_2| \leq \dots$, and $(u_j, u_k)_T = \delta_{jk}$, with δ_{jk} denoting the Kronecker delta. Now let $\phi \in C_0^\infty(\Omega^\#)$, let $f(x) = \phi(x)$ for $x \in \Omega^\#$, $f(x) = 0$ for $x \in \Omega \backslash \Omega^\#$, and define the function

F in Ω by putting $F = f$ if $0 \in \rho(A)$, $F = f - \sum_{j=1}^{n} c_j(f)\Psi_j$ otherwise, where $c_j(f) = (f, \Psi_j)_T/(\Psi_j, \Psi_j)_T$ and the Ψ_j are defined in Theorem 3.2. Then it follows from Theorems 2.4 and 3.4 that $F \in D(A_0)$. We next define the function g on Ω by putting $g(x) = (\omega(x))^{-1}(LF)(x)$ at those points of $\Omega^{\#}$ where $(LF)(x)$ is well defined in the classical sense and putting $g(x) = 0$ at all other points of Ω. Then it is clear that $g \in \mathcal{H}$ and $LF = \omega g$ in $\mathcal{H}$, while this last equation and Theorem 3.4 show that $g \in \mathcal{H}_0$. Hence $A_0 F = T_0 g$ and so $F = Kg$. Thus we conclude from [**154**, Theorem 5, p.410] that

$$F = \sum_{j=1}^{\infty} (f, u_j)_T u_j + h, \tag{4.2}$$

where $h \in N(T_0)$.

Let $u \in \mathcal{H}^{\#} = L^2(\Omega^{\#})$ and let $\|\ \|^{\#}$ denote the norm in $\mathcal{H}^{\#}$. Then given any $\epsilon > 0$, there exists a $\phi \in C_0^{\infty}(\Omega^{\#})$ such that $\|u - \phi\|^{\#} < \epsilon/2$. On the other hand, if $0 \in \rho(A)$ and f denotes the extension of ϕ to Ω defined above, then it follows from (4.2) that

$$\left\| \phi - \sum_{j=1}^{p} (f, u_j)_T u_j \right\|^{\#} \leq \left\| f - \sum_{j=1}^{p} (f, u_j)_T u_j - h \right\| < \epsilon/2$$

if p is sufficiently large. This shows that the eigenvectors of the problem (2.1) are complete in $\mathcal{H}^{\#}$ when $0 \in \rho(A)$, and similar arguments show that this is also the case when $0 \in \sigma(A)$.

Finally, let $u \in \mathcal{H}_{\omega}^{\#} = L^2(\Omega^{\#}; \omega(x)dx)$ and let $\|\ \|_{\omega}^{\#}$ denote the norm in $\mathcal{H}_{\omega}^{\#}$. Then given any $\epsilon > 0$ there exists a $\phi \in C_0^{\infty}(\Omega^{\#})$ such that $\|u - \phi\|_{\omega}^{\#} < \epsilon/2$. On the other hand, if we suppose that $0 \in \rho(A)$, let f denote the extension of ϕ to Ω defined above, and observe that

$$\left\| \phi - \sum_{j=1}^{p} (f, u_j)_T u_j \right\|_{\omega}^{\#} \leq c \left\| f - \sum_{j=1}^{p} (f, u_j)_T u_j - h \right\|$$

for every $p \in \mathbb{N}$, where the constant c does not depend upon p, then it follows from (4.2) that $\|\phi - \sum_{j=1}^{p} (f, u_j)_T u_j\|_{\omega}^{\#} < \epsilon/2$ if p is sufficiently large. Thus we see that the eigenvectors of the problem (2.1) are complete in $\mathcal{H}_{\omega}^{\#}$ when $0 \in \rho(A)$, and similar arguments show that this is also the case when $0 \in \sigma(A)$.

4.3 The system (1.1-4)

We know from the Sturm theory that if $\lambda^{\dagger}$ is an eigenvalue of the system (1.1-4), then

$\lambda^\dagger$ is real (see §1.3) if and only if its second component is real. Hence as a consequence of Theorems 2.5-6, 4.1-2, and Remark 4.1 we now have

Theorem 4.3. The eigenvalues of the system (1.1-4) form a denumerably infinite subset of $\mathbb{R}^2$ having no finite points of accumulation. Moreover, there are at most n^- eigenvalues whose second components are negative. Finally, if $\lambda^\dagger$ is an eigenvalue of the system (1.1-4) and if we put $\phi(x) = \psi^*(x, \lambda^\dagger)$, then $(\phi, \phi)_T > 0$.

Theorem 4.4. The eigenfunctions of the system (1.1-4) are complete in $L^2(\Omega^\#)$ and in $L_2(\Omega^\#; \omega(x)dx)$.

If $0 \in \rho(A)$, then let $\{\lambda(j)\}_1^\infty$ denote an arbitrary enumeration of the eigenvalues of the system (1.1-4), while if $0 \in \sigma(A)$, then let $\{\lambda(j)\}_{n+1}^\infty$ denote an arbitrary enumeration of those eigenvalues of the system (1.1-4) whose second components are not zero (see Theorems 3.1-2). For $j \geq 1$, let

$$\psi_j(x) = \psi^*(x, \lambda(j)) / \left[\int_\Omega \omega(x) \left(\psi^*(x, \lambda(j)) \right)^2 dx \right]^{1/2} .$$

Observe from Theorem 1.2 and Remark 4.1 that $(\psi_j, \psi_k)_T = \delta_{jk}$ for $j, k \geq 1$, where δ_{jk} denotes the Kronecker delta. It is also clear from Theorem 2.6 that the characteristic vectors u_j of the operator K introduced in the proof of Theorem 4.1 can be chosen so that the sequence $\{u_j\}_1^\infty$ is but a rearrangement of the sequence of eigenfunctions of the system (1.1-4), $\{\psi_j(x)\}_p^\infty$, where $p = 1$ if $0 \in \rho(A)$ and $p = n + 1$ otherwise. For the remainder of this chapter it will always be supposed that the u_j have been chosen in this way.

As a consequence of the foregoing results we are now in a position to consider expansion theorems associated with the system (1.1-4). To this end we shall need

Definition 4.2. We say that a series of complex-valued functions all defined on a given set X converges regularly on X if the series of absolute values of these functions converges uniformly on X.

Note that if a series converges regularly, then the series arising from any rearrangement of the terms of the original series also converges regularly and converges uniformly to the same sum function.

Bearing in mind the definitions of §§2.2 and 3.4, we now have

Theorem 4.5. Let f be a function of class $C^{1,1}$ on I^2 which satisfies the boundary conditions (1.2) and (1.4). Suppose also that f vanishes in a relatively open subset of I^2 containing I^2_ω. Then

$$f(x) = \sum_{j=1}^{\infty}(f,\psi_j)_T\psi_j(x) + h(x) \quad \text{for} \quad x \in I^2, \tag{4.3}$$

where the series on the right side of (4.3) converges regularly, and hence uniformly, on I^2, and $h(x)$ is a continuous function on I^2 such that $h \in N(T)$. Finally, if $|I^2_\omega| = 0$, then $h(x) \equiv 0$ and the series on the right side of (4.3) converges uniformly to $f(x)$ on I^2.

Proof. We observe from Theorem 2.4 that $f \in D(A)$. Hence turning to the proof of Theorem 4.2, we may again define F as we did there, but now using the f of the present theorem in place of the f given there, and show us before that $F \in D(A_0)$. Then by arguing as in the proof of Theorem 4.2, we can easily verify that (4.2) remains valid.

Let us now show that

$$\sum_{j=1}^{\infty}\left(\mu_j^{-1}u_j(x)\right)^2 \le c \quad \text{for} \quad x \in I^2, \tag{4.4}$$

where c denotes a positive constant. To this end let k be an arbitrary positive integer, let $\{d_j\}_1^k$ be arbitrary real constants, let $\phi = \sum_{j=1}^k d_j\mu_j u_j$, and put $u = K\phi = \sum_{j=1}^k d_j u_j$. Now observe from Theorems 2.4 and 3.6 that the mapping $A_0^{-1} : R(A_0) \to \mathcal{H}_0$ is bounded and $R(A_0^{-1}) \subset H^2(\Omega)$. Hence it follows from the closed graph theorem (see [**4**, Lemma 13.4, p.210]) that the mapping $A_0^{-1} : R(A_0) \to H^2(\Omega)$ is bounded. Thus we conclude from the Sobolev imbedding theorem [**4**, p.32] that for $x \in I^2$, $|u(x)| \le c_1\|u\|_{2,\Omega} \le c_2\|T_0\phi\| \le c_3\|T^{1/2}\phi\| = c_3\|(PT_0)^{1/2}\phi\| = c_3\left[\sum_{j=1}^k (d_j\mu_j)^2\right]^{1/2}$, where the constants c_j do not depend upon x, k, and the d_j, P is defined in the proof of Theorem 4.1, $T^{1/2}$ and $(PT_0)^{1/2}$ denote the positive square roots of T and PT_0, respectively, and we have used the fact that $\{(PT_0)^{1/2}u_j\}_1^\infty$ is an orthonormal sequence in $\mathcal{H}_0$. Taking $d_j = \mu_j^{-2}u_j(x)$ for $j = 1, \ldots, k$, we obtain $\sum_{j=1}^k\left(\mu_j^{-1}u_j(x)\right)^2 \le c_3^2$, and hence (4.4) now follows from the fact that k is arbitrary.

Finally, observing from the proof of Theorem 4.2 that $F = Kg$, where $g \in \mathcal{H}_0$, and hence that $(f,u_j)_T = (F,u_j)_T = \mu_j^{-1}((PT_0)^{1/2}g, (PT_0)^{1/2}u_j)$, it follows immediately from Cauchy's inequality and (4.4) that the series $\sum_{j=1}^\infty (f,u_j)_T u_j(x)$ converges regularly on I^2. In light of (4.2), the proof of the theorem is complete.

When $|I^2_\omega| > 0$ the expansion (4.3) is not suitable due to the presence of the term $h(x)$. To overcome this deficiency we require

Definition 4.3. Let $p \in \mathbb{N}$. Then we say that the system (1.1-4) satisfies the condition (C_p) if p_r is of class $C^{p,1}$ and A_r, B_r, and q_r are of class $C^{p-1,1}$ in $0 \le x_r \le 1$ for $r = 1,2$.

Recalling that we have written D_r for $\partial/\partial x_r$, we now have

Theorem 4.6. Suppose that the system (1.1-4) satisfies the condition (C_2). Suppose also that f is a function of class $C^{3,1}$ on I^2 with the following properties: (1) f and its partial derivatives up to and including the second order vanish on Γ, (2) $(D_r^3 f)(x)$ vanishes on $x_r = 0$ (resp. $x_r = 1$) if $\alpha_r \neq 0$ (resp. $\beta_r \neq \pi$) for $r = 1,2$, and (3) $f(x)$ vanishes in a relatively open subset of I^2 containing I^2_ω. Then

$$f(x) = \sum_{j=1}^{\infty}(f, \psi_j)_T \psi_j(x) \quad \text{for} \quad x \in I^2, \tag{4.5}$$

where the series on the right side of (4.5) converges regularly and uniformly to $f(x)$ on I^2.

Proof. Let us define the function F on I^2 by putting $F = f$ if $0 \in \rho(A)$ and $F = f - \sum_{j=1}^{n} c_j(f)\Psi_j$ otherwise, where we refer to the proof of Theorem 4.2 for the definitions of the $c_j(f)$ and Ψ_j. Then, as in the proof of Theorem 4.2, we can show that $F \in D(A_0)$. We next define the function $g_1(x)$ in I^2 by putting $g_1(x) = (\omega(x))^{-1}(LF)(x)$ for $x \in I^2 \backslash I^2_\omega, g_1(x) = 0$ for $x \in I^2_\omega$. Then it is clear that $g_1 \in \mathcal{H}$ and $LF = \omega g_1$ in $\mathcal{H}$, while this last equation and Theorem 3.4 show that $g_1 \in \mathcal{H}_0$. Moreover, it follows from the hypotheses of the theorem that $g_1 \in C^{1,1}(I^2)$ and satisfies the boundary conditions (1.2) and (1.4), while g_1 vanishes in a relatively open subset of I^2 containing I^2_ω. In particular, we conclude from Theorem 2.4 that $g_1 \in D(A)$, and hence $g_1 \in D(A_0)$. We now define the function $g_2(x)$ in Ω by putting $g_2(x) = (\omega(x))^{-1}(Lg_1)(x)$ at those points of $\Omega^\#$ where $(Lg_1)(x)$ is well defined in the classical sense and putting $g_2(x) = 0$ at all other points of Ω. Then it is clear that $g_2 \in \mathcal{H}$ and $Lg_1 = \omega g_2$ in $\mathcal{H}$, while this last equation and Theorem 3.4 show that $g_2 \in \mathcal{H}_0$. Thus we have $A_0 g_1 = T_0 g_2, A_0 F = T_0 g_1$, and hence $F = K^2 g_2$. It follows from [**154**, Theorem 5, p.410] that

$$F = \sum_{j=1}^{\infty} (f, u_j)_T u_j \tag{4.6}$$

in $\mathcal{H}_0$. Since $(f, u_j)_T = (F, u_j)_T = \mu_j^{-1}(g_1, u_j)_T$, we may argue as in the proof of Theorem 4.5 to show that the series $\sum_{j=1}^{\infty}(f, u_j)_T u_j(x)$ converges regularly on I^2, and hence the theorem now follows from (4.6).

When $\gamma \geq 0$, a reduction in the hypotheses of Theorem 4.6 is possible.

Theorem 4.7. Suppose that the system (1.1-4) satisfies the condition (C_1) and that $\gamma \geq 0$. Suppose also that f is a function of class $C^{2,1}$ on I^2 with the following properties: (1) f satisfies the boundary conditions (1.2) and (1.4), (2) $(D_r f)(x)$ and $(D_r^2 f)(x)$ vanish on $x_r = 0$ (resp. $x_r = 1$) if $\alpha_r = 0$ (resp. $\beta_r = \pi$) for $r = 1, 2$, and (3) $f(x)$ vanishes in a relatively open subset of I^2 containing I^2_ω. Then the conclusions of Theorem 4.6 remain valid.

Proof. It follows from Theorems 3.6-7 that when V_0, considered only as a vector space, is equipped with the inner product $\langle\ ,\ \rangle = B_0(\ ,\ )$, then it becomes a Hilbert space and that $\|\ \|_0$ and $\|\ \|_{1,\Omega}$ are equivalent norms on V_0, where $\|u\|_0 = \langle u, u\rangle^{1/2}$. Moreover, if we henceforth suppose that V_0 is equipped with the inner product $\langle\ ,\ \rangle$ and if we let $K_0 = K|V_0$, where K is defined in the proof of Theorem 4.1, then it follows from the properties of K cited above and the fact that

$$\langle K_0 u, v\rangle = (T_0 u, v) \tag{4.7}$$

for $u, v \in V_0$, that K_0 is a compact selfadjoint operator in V_0. Note also that the characteristic values and characteristic vectors of K, $\{\mu_j\}_1^\infty$ and $\{u_j\}_1^\infty$, respectively, are precisely those of K_0 and that $\{v_j\}_1^\infty$ is an orthonormal sequence in V_0, where $v_j = \mu_j^{-1/2} u_j$.

Let F be defined as in the proof of Theorem 4.2, except now we use the f of the present theorem in place of the f given there. Then it is clear that $F \in D(A_0)$. Now let $g(x) = (\omega(x))^{-1}(LF)(x)$ for $x \in \Omega^\#$ and $g(x) = 0$ for $x \in \Omega \backslash \Omega^\#$. Then a simple argument involving Definition 2.1 shows that $g \in V$ and $LF = \omega g$ in $\mathcal{H}$, while Theorem 3.4 shows that $g \in V_0$. Hence $A_0 F = T_0 g$, and so $F = K_0 g$. It follows from the spectral theorem and (4.7) that

$$F = \sum_{j=1}^{\infty} \mu_j^{-1} \langle g, v_j \rangle v_j = \sum_{j=1}^{\infty} (f, u_j)_T u_j \text{ in } V_0. \tag{4.8}$$

Since $(f, u_j)_T = (F, u_j)_T = \mu_j^{-1}(g, u_j)_T$, we may argue as we did in the proof of Theorem 4.5 to show that the series $\sum_{j=1}^{\infty}(f, u_j)_T u_j(x)$ converges regularly on I^2, and hence the theorem now follows from (4.8).

Remark 4.3. Theorem 4.7 also remains valid when the condition $\gamma \geq 0$ is replaced by the conditions: (1) $\gamma < 0$ and (2) the inner product $B(\ ,\ )$ is not degenerate on $N(T) \cap V$. Since the proof of this assertion depends upon the theory of Pontrjagin spaces, it will be deferred until Chapter 6 (see Remark 6.14).

4.4. Comments

Turning to §4.2 and fixing our attention firstly upon the situation where L is an ordinary differential operator (so that (2.1) reduces to a weighted Sturm-Liouville problem), Theorem 4.1, at least for the case where the weight function ω is only allowed to vanish on a set of measure zero, has a very long history and goes as far back as Sturm [**148**]. We refer to the papers [**140 - 142**] of Richardson for a further discussion on the early development of this theorem. For the more general case when ω is also allowed to vanish on a set of positive measure, most of the assertions of Theorem 4.1 have been proved by Atkinson [**8**, Chapter 8], under certain restrictions, and by Everitt, Kwong, and Zettl [**67**]. In recent times there has also been some interest in the related problem of ascertaining the asymptotic behaviour of the spectral distribution function and we refer to Jörgens [**114**, §I.5], Gohberg and Krein [**99**, Chapter VI], and Atkinson and Mingarelli [**13**] for relevant investigations. Lastly, for related results concerning singular problems see Kaper, Kwong, and Zettl [**119**]. Fixing our attention next upon the case where L is an elliptic partial differential operator, Theorem 4.1 can also be deduced from the results of Birman and Solomjak [**41 - 43**], where only the case $\gamma > 0$ is considered and the boundary conditions are always taken to be Dirichlet or Neumann (note that these authors consider problems in $\mathbb{R}^k$, $k \in \mathbb{N}$, involving operators not necessarily of the second order). Moreover, the asymptotic behaviour of the spectral distribution function is also investigated in these papers. We remark that in all of the references cited so far in this section in which no restrictions are imposed upon the sign of γ, no bounds for the number of negative eigenvalues are given. Observing that we have defined the eigenvalues and eigenvectors of the problem (2.1) as those of the pencil $A - \lambda T$, a proof of Theorem 4.1 for the case of a general pencil $A - \lambda T$, where A is a selfadjoint operator bounded from below with compact resolvent and T is a bounded selfadjoint operator, has been given by Binding and Seddighi [**39**] under the assumption that $0 \in \rho(A)$ plus an assumption concernng non-degeneracy, and by Binding and Browne [**30**] under the assumption that T is invertible; and referring to the last assertion of the theorem, both these papers give a more precise estimate for the number of negative eigenvalues than we have. Finally, a proof of Theorem 4.1 for the case of a general pencil $A - \lambda T$ has also been given by Weinberger [**153**] under the assumptions that A is a selfadjoint operator with positive lower bound, T is symmetric, $D(T) \supset D(A)$, and $A^{-1}T$ is A-compact.

For the case of ordinary differential operators, the assertion of Theorem 4.2 concerning completeness in $L^2(\Omega; \omega(x)dx)$ is proved in Jörgens [**114**, §I.4] when ω is only allowed to vanish on a set of measure zero, and in Everitt, Kwong, and Zettl [**68**] when this restriction on ω is removed. A proof of Theorem 4.2 for ordinary differential operators is also given in Atkinson [**8**, Chapter 8]; here ω is allowed to vanish on a set of positive measure, but other restrictions are introduced. Lastly, proofs of Theorem 4.2 for the case of a general pencil $A - \lambda T$ are given in [**39**] and [**153**] under conditions already cited.

Turning to §4.3, Theorem 4.3-4 are generalizations of the known results for the special cases treated in [**75,77**], and as in these latter works, the proofs depend upon the theory of compact symmetrisable transformations in Hilbert space. The assertions of Theorems 4.3-4 for the case $\gamma > 0$ can also be deduced from the results of Binding [**17- 19**], Källström and Sleeman [**116, 117**], Sleeman [**144, 145**, and **146**, Chapter 5], and Volkmer [**152**, Chapter 3 and 6], where more general multiparameter systems than considered here are dealt with. Note that all these authors work under the so called left definiteness hypothesis (which, under our assumptions, is the further requirement that $\gamma > 0$ – see [**152**, p.43] for the precise definition) or its generalizations and all consider completeness in function spaces different to ours. More general multiparameter systems then considered here are investigated in Binding and Seddighi [**38**] and Binding [**24**] under assumptions which have already been described in §3.5. Then under their assumptions, and bearing in mind Proposition 3.1, all the assertions of Theorems 4.3-4 can be deduced from the results of each of these papers in combination with the results of Binding and Browne [**30**] and Bognár [**47**, Lemma 3.8, p.35]. Note that the problem (1.1-4) under the right semi–definite assumption of this chapter, but without condition (ii) of Assumption 1.1 (however, with certain other restrictions), falls into a class of non-uniformly right definite multiparameter eigenvalue problems studied by Atkinson [**11**, Chapter 11] and Binding [**22**], and hence one would expect that some of the assertions of this section could be deduced from the results of these authors. Indeed, this is certainly the case. For example, if we suppose that $|I_\omega^2| = 0$, then the initial and final assertions of Theorem

4.3 follow from the results of [**22**], while if we suppose moreover that the system (1.1-4) satisfies the condition (C_2), then the assertion of Theorem 4.4 concerning completeness in $L^2(\Omega^{\#};\omega(x)dx)$ can also be deduced from the results of [**22**] (see Berezanskii [**16**, Theorem 3.2, p.53]). For a further discussion of the eigenvalues of the system (1.1-4), at least when $\gamma > 0$, see Binding and Volkmer [**40**].

Finally, fixing our attention upon the uniform convergence of the eigenfunction expansion, the proof of Theorem 4.5 follows the arguments of [**75, 77**], while the results of Theorems 4.6-7 and Remark 4.3 appear to be new.

CHAPTER 5

The left definite case

5.1. Introduction

In this chapter we shall suppose that ω assumes both positive and negative values in I^2 and that B_0 is positive definite on V_0 (see Theorem 3.7). We have seen that this latter case always occurs when $\gamma \geq 0$ and possibly when $\gamma < 0$, $0 \in \sigma(A)$, and the inner product $(\ ,\ )_T$ is degenerate on N_0. Then as a consequence of the results of Chapters 1–3, we are now in a position to establish some important information concerning the spectral properties of the problem (2.1) and the system (1.1-4). Thus in §5.2 we prove some results concerning the eigenvalues of the problem (2.1) and show that the eigenvectors together with the associated vectors form a complete set in certain function spaces. Analogous results for the system (1.1-4) are proved in §5.3 and in this section we also deal with the uniform convergence of the associated eigenfunction expansion. Finally, in §5.4 we conclude the chapter with some comments.

5.2. The problem (2.1)

We are now going to present our main results concerning the problem (2.1).

Theorem 5.1. The eigenvalues of the problem (2.1) are each of finite algebraic multiplicity and form a denumerably infinite subset of $\mathbb{R}$ having no finite points of accumulation. Moreover, there are infinitely many positive and infinitely many negative eigenvalues and each non-zero eigenvalue is semi-simple.

Proof. We see from Theorem 3.7 that when V_0, considered only as a vector space, is equipped with the inner product $\langle\ ,\ \rangle = B_0(\ ,\ )$, then it becomes a Hilbert space and $\|\ \|_0$ and $\|\ \|_{1,\Omega}$ are equivalent norms on V_0, where $\|u\|_0 = \langle u, u\rangle^{1/2}$. We shall for the remainder of this chapter always suppose, unless otherwise stated, that V_0 is equipped with the inner product $\langle\ ,\ \rangle$. Now observe from Theorem 3.6 that $K = A_0^{-1}T_0$ is a compact operator in $\mathcal{H}_0$, and hence if $K_0 = K|V_0$, then it follows from (4.7) that K_0 is a compact selfajoint operator in V_0. We shall henceforth let P^+ and P^- denote the positive and negative spectral projections, respectively, of K_0 (see the final paragraph of §3.3) and put $\mathcal{L}^\pm = P^\pm V_0$.

We are next going to show that $\dim \mathcal{L}^{\pm} = \infty$. Accordingly, let us consider V only as a vector space, equip it with the inner product $(\ ,\ )_T$, and in this inner product space let $(\dot{+})$ denote the orthogonal direct sum of subspaces of V. Then we observe from the spectral theorem and (4.7) that for $0 \in \rho(A)$, V admits the fundamental decomposition

$$V = N(K_0)(\dot{+})\mathcal{L}^+(\dot{+})\mathcal{L}^- \tag{5.1}$$

with neutral component $N(K_0)$, positive definite component $\mathcal{L}^+$, and negative definite component $\mathcal{L}^-$ (see [**47**, Theorem 5.2, p.89]). On the other hand, if $0 \in \sigma(A)$, then we see from Theorems 3.4-5 that M_0 is a non-degenerate subspace of V, and hence we conclude from [**47**, Corollary 11.8, p.26] that M_0 admits the fundamental decomposition $M_0 = \mathcal{M}^+(\dot{+})\mathcal{M}^-$ with positive definite component $\mathcal{M}^+$ and negative definite component $\mathcal{M}^-$, and where $\mathcal{M}^{\pm}$ are finite dimensional subspaces of V. Thus if $0 \in \sigma(A)$, then it follows that V admits the fundamental decomposition

$$V = N(K_0)(\dot{+})(\mathcal{L}^+(\dot{+})\mathcal{M}^+)(\dot{+})(\mathcal{L}^-(\dot{+})\mathcal{M}^-). \tag{5.2}$$

In light of (5.1-2) we can now employ arguments similar to those used in the proof of Proposition 3.3 to show that if $\mathcal{L}$ is a positive definite subspace of V, then $\mathcal{L}$ is algebraically isomorphic to a subspace of $\mathcal{L}^+$ if $0 \in \rho(A)$ and to a subspace of $\mathcal{L}^+(\dot{+})\mathcal{M}^+$ otherwise. But it is clear from the definition of ω that V contains positive definite subspaces of arbitrary large dimension, and hence we conclude that $\dim \mathcal{L}^+ = \infty$. Similarly we can show that $\dim \mathcal{L}^- = \infty$.

As a consequence of the foregoing results we see that the non-zero eigenvalues of K_0 are real, with each being semi-simple and of finite multiplicity, and that there are infinitely many positive eigenvalues as well as infinitely many negative ones. On the other hand, we can argue as we did in the proof of Theorem 4.1 to show that the characteristic values of K_0 are precisely the non-zero eigenvalues of the problem (2.1), and that for λ a non-zero eigenvalue of (2.1), $N_\lambda = M_\lambda = \mathcal{G}_{1/\lambda}$, where $\mathcal{G}_{1/\lambda}$ denotes the eigenspace of K_0 corresponding to the eigenvalue $1/\lambda$. The assertions of the theorem now follow from these results and Theorems 3.4-5.

<u>Remark 5.1.</u> It follows from Theorems 3.1-2, 3.4-5, and 5.1 that the eigenvalues of the problem (2.1) are all non-zero if $\gamma > 0$ and all semi-simple if $\gamma = 0$ and the inner product

$(\, , \,)_T$ is not degenerate on N_0. On the other hand if $\gamma < 0$ or $\gamma = 0$ and the inner product $(\, , \,)_T$ is degenerate on N_0, then the eigenvalue $\lambda = 0$ of (2.1) is never semi-simple and in the former (resp. latter) case its algebraic multiplicity does not exceed $2(n + n^-)$ (resp. $2n$).

Remark 5.2. It follows from (4.7) and the proof of Theorem 5.1 that if $\lambda \neq 0$ is an eigenvalue of the problem (2.1) and u a corresponding eigenvector, then $(u,u)_T \neq 0$ and $sgn(u,u)_T = sgn\lambda$.

Recalling the definition of $\Omega^\#$ given in Chapter 4 and bearing in mind Remark 4.2, we have next

Theorem 5.2. The eigenvectors of the problem (2.1) together with their associated vectors, if any, are complete in $L^2(\Omega^\#)$ and in $L^2(\Omega^\#; |w(x)|dx)$.

Proof. Turning again to the proof of Theorem 5.1, we see that the characteristic values and characteristic vectors of K_0 may be arranged into sequences $\{\mu_j\}_1^\infty$ and $\{u_j\}_1^\infty$, respectively, where each characteristic value is repeated according to its multiplicity, $0 < |\mu_1| \leq |\mu_2| \leq \ldots$, $(u_j, u_k)_T = \delta_{jk} sgn \mu_j$, $\langle u_j, u_k \rangle = \delta_{jk} |\mu_j|$, and δ_{jk} is the Kronecker delta. Putting $v_j = |\mu_j|^{-1/2} u_j$ for $j \geq 1$, it follows from the spectral theorem that the v_j form an orthonormal basis of $\mathcal{L}^+ \dot{+} \mathcal{L}^-$, and hence a basis of $\mathcal{L}^+ \dot{+} \mathcal{L}^-$ with respect to the topology induced on this space by the norm $\| \; \|$.

Let $u \in \mathcal{H}^\# = L^2(\Omega^\#)$ and let $f(x) = u(x)$ for $x \in \Omega^\#$, $f(x) = 0$ for $x \in \Omega \backslash \Omega^\#$. Then it follows from Theorems 3.4-5 that when $0 \in \sigma(A)$, $f = f_0 + f_1$, where $f_0 \in \mathcal{H}_0$ and $f_1 \in M_0$. If $0 \in \rho(A)$, then we will put $f_0 = f$, $f_1 = 0$. Now given any $\epsilon > 0$, we know from Theorem 3.6 that there is a $g \in V_0$ such that $\|f_0 - g\| < \epsilon$. Hence, since $g = g_0 + P^+ g + P^- g$, where $g_0 \in N(K_0) \subset N(T)$, we see that if $\| \; \|^\#$ denotes the norm in $\mathcal{H}^\#$, then $\|u - f_1 - P^+ g - P^- g\|^\# = \|u - f_1 - g\|^\# \leq \|f - f_1 - g\| < \epsilon$, and the first assertion of the theorem now follows.

Finally, let $u \in \mathcal{H}_\omega^\# = L^2(\Omega^\#; |\omega(x)|dx)$ and let ϵ be any positive number. Then there is a $\phi \in C_0^\infty(\Omega^\#)$ such that $\|u - \phi\|_\omega^\# < \epsilon$, where $\| \; \|_\omega^\#$ denote the norm in $\mathcal{H}_\omega^\#$. Since $\|y\|_\omega^\# \leq c\|y\|^\#$ for $y \in \mathcal{H}^\#$, where c denotes a positive constant, it follows from what has already been proved that there is a finite linear combination of eigenvectors and associated vectors of the problem (2.1), say χ, such that $\|\phi - \chi\|_\omega^\# < \epsilon$. This shows that the eigenvectors and associated vectors of the problem (2.1) are complete in $\mathcal{H}_\omega^\#$ and completes the proof of the theorem.

5.3. The system (1.1-4)

We have already mentioned that an eigenvalue of the system (1.1-4) is real if and only if its second component is real. Hence it follows from Theorems 2.5-6, 5.1, and Remark 5.2 that

Theorem 5.3. The eigenvalues of the system (1.1-4) form a denumerably infinite subset of $\mathbb{R}^2$ having no finite points of accumulation. Moreover, there are infinitely many eigenvalues whose second components are positive and infinitely many with negative second components. Finally if $\lambda^\dagger = (\lambda_1^\dagger, \lambda_2^\dagger)$ is an eigenvalue of the system (1.1-4), if $\lambda_2^\dagger \neq 0$, and if $\phi(x) = \psi^\star(x, \lambda^\dagger)$, then $(\phi, \phi)_T \neq 0$.

We have seen in §5.2 that if λ is an eigenvalue of the problem (2.1) and λ is not semi-simple, then $\lambda = 0$. We have also shown in Theorem 2.6 that each eigenspace of the problem (2.1) has a basis consisting of eigenfunctions of the system (1.1-4). Hence in light of the definitions given in the statements following (2.4), we shall from now on, when dealing with the system (1.1-4), refer to the χ_{jk} of Theorem 3.5 as functions associated with the eigenfunctions Ψ_j, $j = 1, \ldots, m$, of (1.1-4) (see Theorem 3.2). Then as a consequence of Theorem 5.2 we have

Theorem 5.4. The eigenfunctions of the system (1.1-4) together with their associated functions, if any, are complete in $L^2(\Omega^\#)$ and in $L^2(\Omega^\#; |w(x)|dx)$.

If $\gamma > 0$, then let $\{\lambda(j)\}_1^\infty$ denote an arbitrary enumeration of the eigenvalues of the system (1.1-4), while if $\gamma \leq 0$, then let $\{\lambda(j)\}_{n+1}^\infty$ denote an arbitrary enumeration of those eigenvalues of the system (1.1-4) whose second components are not zero (see Theorems 3.1-2). If $\gamma > 0$ or if $\gamma = 0$ and the inner product $(\ ,\ )_T$ is not degenerate on N_0, then let

$$\psi_j(x) = \psi^\star(x, \lambda(j)) \Big/ \left| \int_\Omega w(x) \Big(\psi^\star(x, \lambda(j)) \Big)^2 dx \right|^{1/2} \tag{5.3}$$

for $j \geq 1$, while if $\gamma < 0$ or if $\gamma = 0$ and the inner product $(\ ,\ )_T$ is degenerate on N_0, then let $\psi_j(x) = \Psi_j(x)$ for $j = 1, \ldots, m$ and let $\psi_j(x)$ be defined by (5.3) for $j > m$. Observe from Theorems 1.2 and 3.2 that $(\psi_j, \psi_k)_T = 0$ if $j \neq k$, while $(\psi_j, \psi_j)_T = \rho_j = \pm 1$ for: (1) $j \geq 1$ if $\gamma > 0$ or if $\gamma = 0$ and the inner product $(\ ,\ )_T$ is not degenerate on N_0, and (2) $j > m$ if $\gamma < 0$ or if $\gamma = 0$ and the inner product $(\ ,\ )_T$ is degenerate on N_0. It is also clear from Theorem 2.6 that the characteristic vectors u_j of the operator K_0 (see the proof of Theorem 5.2) can be chosen so that the sequence $\{u_j\}_1^\infty$ is but a rearrangement of the sequence of eigenfunctions of the system (1.1-4), $\{\psi_j\}_p^\infty$, where $p = 1$ if $\gamma > 0$ and $p = n + 1$ otherwise. For the remainder of this chapter it will always be supposed that the u_j have been chosen in this way.

Theorem 5.5. In order that the eigenfunctions of the system (1.1-4) be complete in $L^2(\Omega^\#)$ (resp. $L^2(\Omega^\#; |\omega(x)|dx)$) it is necessary and sufficient that either $\gamma > 0$ or $\gamma = 0$ and the inner product $(\ ,\ )_T$ is not degenerate on N_0.

Remark 5.3. In view of the results of §5.2 and Theorems 3.2 and 3.4, we see that the theorem can be equivalently stated as: In order that the eigenfunctions of the system (1.1-4) be complete in $L^2(\Omega^\#)$ (resp. $L^2(\Omega^\#; |\omega(x)|dx)$) it is necessary and sufficient that $(\psi_j, \psi_j)_T \neq 0$ for $j \geq 1$.

Proof of Theorem 5.5. If $\gamma > 0$ or if $\gamma = 0$ and the inner product $(\ ,\ )_T$ is not degenerate on N_0, then the completeness of the eigenfunctions of the system (1.1-4) in the spaces concerned has already been established (see Theorems 3.1 and 3.4). Suppose conversely that the eigenfunctions of (1.1-4) are complete in either of the spaces concerned and suppose also that either $\gamma < 0$ or $\gamma = 0$ and the inner product $(\ ,\ )_T$ is degenerate on N_0. Fixing our attention upon Theorem 3.5, let us now fix a $j, 1 \leq j \leq m$, and put $\chi(x) = \chi_{j,p_j}(x)$. Then there is a sequence $\{\chi_k\}_1^\infty$ of finite linear combinations of eigenfunctions of the system (1.1-4) such that $\chi_k|\Omega^\# \to \chi|\Omega^\#$ in either $L^2(\Omega^\#)$ or $L^2(\Omega^\#; |\omega(x)|dx)$, depending upon the space being considered. Hence in view of Theorems 1.2, 3.2, and 3.5, we arrive at the contradiction that $(\Psi_j, \chi)_T = \lim_{k\to\infty}(\Psi_j, \chi_k)_T = 0$. This contradiction completes the proof of the theorem.

We are now going to state our main results concerning the eigenfunction expansion associated with the system (1.1-4). Accordingly, recalling Definitions 4.2-3 as well as the definition of ρ_j given above, we have firstly

Theorem 5.6. Suppose that the system (1.1-4) satisfies the condition (C_1) and that either $\gamma > 0$ or $\gamma = 0$ and the inner product $(\ ,\ )_T$ is not degenerate on N_0. Let f satisfy the hypotheses given in Theorem 4.7. Then

$$f(x) = \sum_{j=1}^{\infty} \rho_j (f, \psi_j)_T \psi_j(x) \quad \text{for} \quad x \in I^2, \tag{5.4}$$

where the series on the right side of (5.4) converges regularly and uniformly to $f(x)$ on I^2.

Turning next to our second result and referring to Theorem 3.5 for the definitions of the χ_{j1} and to (3.15) for the definitions of the $c_j(f)$, $c_{j1}(f)$, we have

Theorem 5.7. Suppose that the system (1.1-4) satisfies the condition (C_1), that $\gamma = 0$, and that the inner product $(\ ,\ )_T$ is degenerate on N_0. Let f satisfy the hypotheses given in Theorem 4.7. Then

$$f(x) = \sum_{j=1}^{m} c_j(f)\psi_j(x) + \sum_{j=1}^{m} c_{j1}(f)\chi_{j1}(x) + \sum_{j=m+1}^{\infty} \rho_j(f,\psi_j)_T\psi_j(x) \quad \text{for } x \in I^2, \tag{5.5}$$

where the infinite series on the right side of (5.5) converges regularly, and hence uniformly, on I^2.

Turning to our final result and referring to Theorem 3.5 for the definitions of the χ_{jk} and to (3.17) for the definitions of the $c_j(f), c_{jk}(f)$, we have

Theorem 5.8. Suppose that the system (1.1-4) satisfies the condition (C_1) and that $\gamma < 0$. Let f satisfy the hypothesis given in Theorem 4.7. Then

$$f(x) = \sum_{j=1}^{m} c_j(f)\psi_j(x) + \sum_{j=1}^{m}\sum_{k=1}^{p_j} c_{jk}(f)\chi_{jk}(x) + \sum_{j=m+1}^{\infty} \rho_j(f,\psi_j)_T\psi_j(x) \quad \text{for } x \in I^2, \tag{5.6}$$

where the infinite series on the right side of (5.6) converges regularly, and hence uniformly, on I^2.

We shall for brevity only prove Theorem 5.8; the proofs of Theorems 5.6–7 are similar.

Proof of Theorem 5.8. To begin with, we observe from the definitions of the terms involved (see the proof of Theorem 3.5) that the ψ_j and χ_{jk} are of class $C^{2,1}$ on I^2. Moreover, we observe from Theorem 2.4 that $f \in D(A)$, and hence if we let

$$F(x) = f(x) - \sum_{j=1}^{m} c_j(f)\psi_j(x) - \sum_{j=1}^{m}\sum_{k=1}^{p_j} c_{jk}(f)\chi_{jk}(x) - \\ - \sum_{j=m+1}^{n} \rho_j(f,\psi_j)_T\psi_j(x) \text{ for } x \in I^2,$$

where the last summation does not appear if $m = n$, then it follows from the definitions of the $c_j(f)$ and $c_{jk}(f)$ that $F \in D(A_0)$. Now let $g(x) = (\omega(x))^{-1}(LF)(x)$ for $x \in \Omega^\#$ and $g(x) = -\sum_{j=1}^{m}\sum_{k=1}^{p_j} c_{jk}(f)\chi_{j,k-1}(x)$ for $x \in \Omega\backslash\Omega^\#$, where $\chi_{j0} = \psi_j$. Then a simple argument involving Definition 2.1 and Theorem 3.5 show that $g \in V$ and $LF = \omega g$ in $\mathcal{H}$, while this last equation and Theorem 3.5 show that $g \in V_0$. Hence $A_0F = T_0g$ and so $F = K_0g$. Referring to the proof of Theorem 5.2 for terminology, it now follows from the spectral theorem and (4.7) that

$$F = \sum_{j=1}^{\infty} \mu_j^{-1}\langle g, v_j\rangle v_j = \sum_{j=1}^{\infty} sgn\mu_j(f,u_j)_T u_j \text{ in } V_0. \tag{5.7}$$

We are now going to show that

$$\sum_{j=1}^{\infty} \left(\mu_j^{-1}v_j(x)\right)^2 \le c \text{ for } x \in I^2, \tag{5.8}$$

where c denotes a positive constant. Accordingly, for $k \in \mathbb{N}$ let $\{d_j\}_1^k$ be arbitrary real constants, let $\phi = \sum_{j=1}^{k} d_j\mu_j v_j$, and put $v = K_0\phi = \sum_{j=1}^{k} d_j v_j$. Then we may argue as we did in the proof of Theorem 4.5 to show that the mapping $A_0^{-1} : R(A_0) \to H^2(\Omega)$ is bounded, and hence it follows from the Sobolev imbedding theorem that for $x \in I^2$,

$$|v(x)| \le c_1\|v\|_{2,\Omega} \le c_2\|T_0\phi\| \le c_3\|\phi\|_{1,\Omega} \le c_4\|\phi\|_0 = c_4\left(\sum_{j=1}^{k}(d_j\mu_j)^2\right)^{1/2},$$

where the constants c_r do not depend upon x, k, and the d_j, while $\| \ \|_0$ is defined in the proof of Theorem 5.1. Taking $d_j = \mu_j^{-2}v_j(x)$ for $j = 1, \ldots, k$, we obtain $\sum_{j=1}^{k}\left(\mu_j^{-1}v_j(x)\right)^2 \le c_4^2$, and hence (5.8) now follows from the fact that k is arbitrary.

Finally, observing that $sgn\mu_j(f,u_j)_T u_j = \mu_j^{-1}\langle g, v_j\rangle v_j$, we conclude from (5.8) that the series $\sum_{j=1}^{\infty} sgn\mu_j(f,u_j)_T u_j(x)$ converges regularly on I^2. Hence, bearing in mind that $(u_j,u_j)_T = sgn\mu_j$, the theorem now follows from (5.7).

5.4 Comments

The title of this chapter does not conform to standard usage in that the left definite nomenclature is usually reserved for the case $\gamma > 0$. However, since we have reduced (2.1) as a problem in $\mathcal{H}$, to a left definite problem in $\mathcal{H}_0$ (see Theorem 3.7), we have felt justified in employing our terminology.

Turning now to §5.2 and fixing our attention for the moment upon the situation where L is an ordinary differential operator (so that (2.1) reduced to a weighted Sturm–Liouville problem), Theorems 5.1–2 seem to have been firstly proved by Hilbert [**106**] under the hypothesis that $\gamma > 0$ plus certain other restrictions on the weight function ω. For a further discussion concerning the early development of Theorem 5.1 see Richardson [**140 - 142**], while for more recent proofs of some or all of the assertions of these theorems under various restrictions see Beals [**15**], Faierman and Roach [**88**], and Kaper et al. [**118**]. Referring again to Theorem 5.1, there has also been some recent interest in the related problem of ascertaining the asymptotic behaviour of the distribution functions of the positive and negative eigenvalues; we refer to Atkinson and Mingarelli [**13**] and Fleckinger and Lapidus [**94**] for relevant investigations. Finally, for relevant results concerning both regular and singular problems see Ćurgus and Langer [**59**], Daho and Langer [**60**], and Kaper, Kwong, and Zettl [**119**].

Fixing our attention next upon the case where L is an elliptic partial differential operator, the earliest proofs of some or all of the assertions of Theorems 5.1–2 seem to be due to Hilbert [**106**] and Hölmgren [**108**], where in both works it is supposed that $\gamma > 0$. Hölmgren actually proved the existence of infinitely many positive and negative eigenvalues for the Laplace operator under Dirichlet boundary conditions, and the asymptotic distribution of these eigenvalues was then established by Pleijel [**138**]. The nature of the spectrum for a more general operator L, but again with the restriction $\gamma > 0$, was also investigated by Pleijel [**139**]. Theorem 5.1 can also be deduced from the results of Birman and Solomjak [**41 - 43**], Fleckinger and Lapidus [**94**], and Hess [**104**], where only the case $\gamma > 0$ is considered and the boundary conditions are always taken to be Dirichlet or Neumann (note that these authors consider problems in $\mathbb{R}^k$, $k \in \mathbb{N}$, involving operators not necessarily of the second order). Moreover, the asymptotic behaviour of the distribution functions of the positive and negative eigenvalues is also investigated in all but the last of these works. The proofs of Theorems 5.1–2 given here follow closely the arguments of [**85**, §3], and for further related results we refer to Beals [**15**] and Faierman and Roach [**91**]. Lastly, proofs of Theorems 5.1–2 for the case of a general pencil $A - \lambda T$ are given in Binding and Seddighi [**39**] and Weinberger [**153**] under conditions already described in §4.4. For further related results for pencils see Binding and Browne [**30**].

Turning next to §5.3, Theorems 5.3–5 are generalizations of the known results for the special cases treated in [**75 - 79**], [**82, 83**], and [**90**], where in all but the last of these references, proofs were established (unlike here) by appealing to the theory of compact symmetrisable transformations in Hilbert space. The proof of Theorem 5.3 given here, which was established via a study of the elliptic boundary value problem (2.1), follows the ideas sketched out in [**81**]. For the analogue of Theorem 5.4 for the case of partial-range eigenfunction expansions see Faierman and Roach [**89**]. When $\gamma > 0$, all the assertions of Theorem 5.3, apart from the second, as well as those of Theorem 5.4 can also be deduced from the results of Binding [**17 - 19**], Källström and Sleeman [**116, 117**], Sleeman [**144, 145**, and **146**, Chapter 5], and Volkmer [**152**, Chapters 3 and 6], where more general multiparameter systems then considered here are dealt with. As we remarked in §4.4, all these authors work under the left definite assumption or its generalizations and all consider completeness in function spaces different to ours. More general multiparameter systems than considered here have also been investigated by Binding and Seddighi [**38**] and Binding [**24**] under assumptions which have already been described in §3.5. Then under their assumptions, all the assertions of Theorem 5.3, apart from the second, as well as those of Theorem 5.4 can be deduced from the results of each of these papers in combination with the results of [**47**, Lemma 3.8, p.35]. Finally, fixing our attention upon the uniform convergence of the eigenfunction expansion dealt with in Theorems 5.6–8, the proof of Theorem 5.8 given here is, apart from certain modifications, taken from Faierman and Roach [**90**, Theorem 6.1].

CHAPTER 6

The non-definite case

6.1. Introduction

In this chapter we shall suppose that $\omega(x)$ assumes both positive and negative values on I^2 and that B_0 is indefinite on V_0. We see from Theorem 3.7 that this latter case always occurs when $\gamma < 0$ and either $0 \in \rho(A)$ or $0 \in \sigma(A)$ and the inner product $(\ ,\ )_T$ is not degenerate on N_0, and possibly when $\gamma < 0$, $0 \in \sigma(A)$, and the inner product $(\ ,\ )_T$ is degenerate on N_0. Then as a consequence of the foregoing results we are now in a position to establish some important information concerning the problem (2.1) and the system (1.1-4). Since much of our work will depend upon the theory of inner product spaces, we introduce in §6.2 some further concepts from this theory which complement those of §3.2. These concepts are then used in §6.3 to show that when V_0 is equipped with the inner product $B_0(\ ,\ )$, it becomes a Pontrjagin space. Hence we are able in §6.4 to apply the Pontrjagin space theory in full force in order to establish our main results concerning the eigenvalues of the problem (2.1) and to show that the eigenvectors together with the associated vectors form a complete set in certain function spaces. The results of §6.4 are then used in §6.5 in order to derive information concerning the spectral properties of the system (1.1-4). In §6.6 we deal with the problem of ascertaining how the principal subspaces of the problem (2.1) corresponding to non-semi-simple eigenvalues are related to the system (1.1-4), while §6.7 is devoted to the study of the uniform convergence of the eigenfunction expansion associated with (1.1-4). Finally, in §6.8, we conclude the chapter with some comments.

6.2. Further facts from the theory of inner product spaces

We return to our discussion of §3.2 concerning the inner product space X, with inner product $[\ ,\ ]$, and say that two subspaces X_1 and X_2 of X form a dual pair if there is no non-zero vector of X_r which is orthogonal to all of X_{3-r} for $r = 1, 2$. We also say that a subspace X_1 of X is maximal positive if X_1 is positive and is not a proper subspace of any other positive subspace of X. An analogous definition holds for maximal negative subspaces of X. A topology τ on X is called a partial majorant if: (1) τ is locally convex, and (2) for any fixed $v \in X$ the function $f_v(u) = [u, v]$ is τ-continuous on X. A topology τ on X is said to be admissible if: (1) τ is a partial majorant on X, and (2) for any τ-continuous linear functional f on X there is a vector $v_0 \in X$ such that $f(u) = [u, v_0]$

for $u \in X$. A topology τ on X is called a majorant if: (1) τ is locally convex, and (2) the inner product $[\ ,\ ]$ is jointly τ-continuous. Lastly, a majorant τ on X is said to be minimal if there is no other majorant of X which is weaker than τ.

Suppose next that X is a decomposable, non-degenerate inner product space and that X^+ and X^- are the positive definite and negative definite components, respectively, corresponding to a fundamental decomposition of X. We define two projectors P^+, P^- in X by the relations $P^+X = X^+, P^-X = X^-$, in other words, we set $P^+x = x^+, P^-x = x^-$ for $x \in X$, where $x = x^+ + x^-(x^+ \in X^+, x^- \in X^-)$ is the decomposition of the vector x corresponding to the fundamental decomposition of X in question. P^+ and P^- are called fundamental projectors and $J = P^+ - P^-$ is called the fundamental symmetry belonging to the fundamental decomposition of X in question. When X, considered only as a vector space, is equipped with the J-inner product $[\ ,\ ]_J = [J.,.]$, then it becomes a positive definite inner product space and $\|x\|_J = [x,x]_J^{1/2}$ is a norm on X which is called the J-norm. If τ_J denotes the topology on X induced by the J-norm, then τ_J is a majorant and X^+ and X^- are τ_J-closed. If X^+ and X^- are complete with respect to the J-norm, then X is called a Krein space, and in this case X is also a Hilbert space with respect to the J-inner product. If X is a Krein space, then all the J-norms arising from all possible fundamental decompositions of X are equivalent, and the Hilbert topology which they define, τ_J, is called the strong toplogy of X. If X is a Krein space and $\kappa^\pm = \dim X^\pm$, where dim denotes the τ_J-dimension of the space concerned, then the cardinal number κ^+ (resp. κ^-) is the same for every fundamental decomposition of X and is called the rank of positivity (resp. negativity) of X. The cardinal number $\kappa = \min\{\kappa^+, \kappa^-\}$ is called the rank of indefiniteness of X. A Krein space with a finite rank of indefiniteness κ is called a Pontrjagin space and κ is called the index of this space.

6.3. The subspace V_0

In this section we are going to derive some information concerning the subspace V_0 of V defined in the statements preceding Theorem 3.6 which we require for later use. Accordingly, we see from Theorem 3.6 that V_0 is a Hilbert space with respect to the inner product $(\ ,\ )_{1,\Omega}$, while from Theorem 3.7 we see that when V_0 is considered only as a vector space and is equipped with the inner product $\langle\ ,\ \rangle = B_0(\ ,\ )$, then it becomes an indefinite inner product space. Moreover, we know from Proposition 2.1 that $|\langle u,v\rangle| \leq c\|u\|_{1,\Omega}\|v\|_{1,\Omega}$ for $u, v \in V_0$, where c denotes a positive constant. Hence it follows that there exists a bounded selfadjoint operator G in V_0 such that

$$\langle u,v\rangle = (Gu,v)_{1,\Omega} \quad \text{for } u,v \in V_0. \tag{6.1}$$

We call G the Gram operator of $\langle\ ,\rangle$ with respect to $(\ ,\ )_{1,\Omega}$ [**47**, p.89]. Now let P_+ and P_- denote the positive and negative spectral projections of G, respectively (see the final paragraph of §3.3), and put $\mathcal{L}_\pm = P_\pm V_0$. Then $\mathcal{L}_+$ and $\mathcal{L}_-$ are Hilbert spaces with respect to the inner product $(\ ,\ )_{1,\Omega}$, and when considered as subspaces of the indefinite inner product space V_0 (i.e., when the vector space V_0 is equipped with the inner product $\langle\ ,\ \rangle$), $\mathcal{L}_+$ is positive definite and $\mathcal{L}_-$ is negative definite. Hence if κ denotes the Hilbert space dimension of $\mathcal{L}_-$ and if we refer to the remarks preceding Proposition 3.3 for terminology, then we have

Proposition 6.1. V_0, when equipped with the inner product $\langle\ ,\ \rangle$, is a Pontrjagin space of index κ, where $0 < \kappa \leq n^-$.

Proof. Firstly we remark that the proposition asserts that V_0 is a Krein space with finite rank of indefiniteness κ and it will be shown below that κ is precisely the rank of negativity of V_0. Passing now to the proof of the proposition, let us firstly show that $0 \in \rho(G)$, or equivalently, that G has a bounded inverse. Indeed, if G does not have a bounded inverse, then it follows from Proposition 2.2 and the fact that $D(A_0^\dagger)$ is a core of B_0 (see Theorem 3.7 and [**120**, Theorem 2.1, p.322]) that there exists the sequence $\{u_j\}$ in $D(A_0^\dagger)$ with $\|u_j\|_{1,\Omega} = 1$ such that $\|Gu_j\|_{1,\Omega} \to 0$ as $j \to \infty$. Since $(Gu_j, v)_{1,\Omega} = \langle u_j, v\rangle = (u_j, A_0^\dagger v)$ for $v \in D(A_0^\dagger)$, it follows from Rellich's theorem [**4**, p.30] and Theorem 3.7 that $\{u_j\}$ has a subsequence which converges to zero in $\mathcal{H}_0$. Consequently, if we denote this subsequence again by $\{u_j\}$ and make use of the fact that B_0 is coercive over V_0 (see (2.2) and Theorem 3.7), then we arrive at the contradiction that $\|u_j\|_{1,\Omega} \to 0$ as $j \to \infty$.

Thus it follows from [**47**, Theorem 1.3, p.101] that V_0 is a Krein space and admits a fundamental decomposition with positive definite component $\mathcal{L}_+$ and negative definite component $\mathcal{L}_-$. Moreover, we see from Theorem 3.7 that $A_0^\dagger$ has a negative lower bound, which implies that $\mathcal{L}_- \neq 0$, and hence $\kappa > 0$. On the other hand we see from Proposition 3.3 that $\kappa \leq n^-$, while the assumption that $\mathcal{L}_+$ is finite dimensional leads to the conclusion that V_0, and hence $\mathcal{H}_0$, is finite dimensional, in view of Theorem 3.6. Hence it follows from Theorems 3.1, 3.4-5 that if $\mathcal{L}_+$ is finite dimensional, then so is $\mathcal{H}$, which is not possible. Thus $\mathcal{L}_+$ cannot be finite dimensional and this completes the proof of the proposition.

Remark 6.1. The bound $\kappa \leq n^-$ of Proposition 6.1 can be further refined. Indeed, by appealing to the proof of Proposition 3.3 (suitably modified for the case $0 \in \rho(A)$),

Theorem 3.6, and [**47**, Corollary 7.4 and Remark 7.5, p.94], it is not difficult to verify that we actually have $\kappa = n^-$ for either of the cases $0 \in \rho(A)$ or $0 \in \sigma(A)$ and the inner product $(\ , \ )_T$ is not degenerate on N_0.

Proposition 6.2. The strong topology of the Pontrjagin space V_0 coincides with the Hilbert topology in V_0 induced by the norm $\| \ \|_{1,\Omega}$.

Proof. If τ denotes the Hilbert topology in V_0 induced by the norm $\| \ \|_{1,\Omega}$, then it follows from (6.1) that τ is both a partial majorant and a majorant on the Pontrjagin space V_0. Moreover, if f is a τ-continuous linear functional on V_0, then there is a $v_0 \in V_0$ such that $f(u) = (u, v_0)_{1,\Omega} = \langle u, G^{-1}v_0\rangle$ for every $u \in V_0$, in light of (6.1). Thus τ is an admissable topology on the Pontrjagin space V_0, and hence it follows from [**47**, Theorem 4.5, p.87] that τ is a minimal majorant. The assertion of the proposition now follows from [**47**, Theorem 6.1, p.90].

6.4. The problem (2.1)

We observe from Theorem 3.6 that $K = A_0^{-1}T_0$ is a compact operator in $\mathcal{H}_0$, and hence if $K_0 = K|V_0$, then it follows from (4.7) and Proposition 6.2 that K_0 is a compact selfadjoint operator in the Pontrjagin space V_0 (here compactness is meant with respect to the strong topology of V_0 and selfadjointness means that $\langle K_0u, v\rangle = \langle u, K_0v\rangle$ for $u, v \in V_0$ – see [**47**, pp.121 and 133]). We are now going to use these facts to derive some information concerning the spectral properties of the problem (2.1).

We have seen in §6.3 that when V_0, considered only as a vector space, is equipped with the inner product $\langle \ , \ \rangle$, then it becomes an indefinite inner product space, and is in fact a Pontrjagin space. On the other hand, we know from Corollary 3.1 that when V_0, again considered only as a vector space, is equipped with the inner product $(\ , \ )_T$, then it also becomes an indefinite inner product space.

Terminology. In the sequel, when we refer to V_0 as a Pontrjagin space, then this will always be meant with respect to the inner product $\langle \ , \ \rangle$, while if we refer to V_0 (resp. V) as an indefinite inner product space, then this will always be meant with respect to the inner product $(\ , \ )_T$.

Remark 6.2. We have seen above that K_0 is a selfadjoint operator in the Pontrjagin space V_0. It also follows from (4.7) that K_0 is a selfadjoint operator in V_0 when this space is considered as an indefinite inner product space in the sense just described.

If we now consider the principal subspaces of the problem (2.1) as subspaces of the indefinite inner product space V, then we have

Theorem 6.1. The eigenvalues of the problem (2.1) are each of finite algebraic multiplicity and form a denumerably infinite subset of $\mathbb{C}$ having no finite points of accumulation. Moreover, if λ and μ are eigenvalues satisfying $\lambda \neq \bar{\mu}$, then M_λ is orthogonal to M_μ, and hence the principal subspace corresponding to a non-real eigenvalue is neutral. Also if λ is an eigenvalue, then so is $\bar{\lambda}$ and M_λ and $M_{\bar{\lambda}}$ form a dual pair; thus if λ is real, then M_λ is non-degenerate. Finally, the eigenspaces corresponding to the distinct eigenvalues of the problem (2.1) are definite, with the possible exception of at most $2\kappa + 1$ of them, and hence the eigenvalues of (2.1) are real and semi-simple, with the possible exception of at most $2\kappa + 1$ of them.

Proof. Let us turn again to the operator K_0 acting in the Pontrjagin space V_0 and henceforth agree to denote the principal subspace of K_0 corresponding to the eigenvalue λ by $\mathcal{G}_\lambda$. Then $\dim \mathcal{G}_\lambda < \infty$ if $\lambda \neq 0$. We are now going to show that the characteristic values of K_0 are precisely the non-zero eigenvalues of the problem (2.1) and that for each characteristic value λ we have $\mathcal{G}_{1/\lambda} = M_\lambda$. Accordingly, let us firstly suppose that $\lambda \neq 0$ is an eigenvalue of the problem (2.1). Then we see from Theorem 3.8 that $M_\lambda \subset V_0$, while it is a simple matter to show that λ is a characteristic value of K_0 and $N_\lambda \subset N(I - \lambda K_0) \subset \mathcal{G}_{1/\lambda}$. Hence if λ is semi-simple, then $M_\lambda \subset \mathcal{G}_{1/\lambda}$. Now suppose that λ is not semi-simple and $u \in M_\lambda, u \notin N_\lambda$. Then there exist the vectors $\{u_k\}_0^{p-1}$ in M_λ, where $p > 1, u_{p-1} = u$, and $u_0 \neq 0$, such that $(A - \lambda T)u_k = Tu_{k-1}$, where $u_{-1} = 0$. If we now define the vectors $\{z_k\}_0^{p-1}$ by putting $\nu = -\lambda^{-2}$ and

$$
\begin{aligned}
u_k &= \nu^k z_k \quad \text{for} \quad k = 0, 1, \\
u_k &= \nu^k z_k + \sum_{r=1}^{k-1} \alpha_{kr} z_r \quad \text{for} \quad 2 \le k \le p-1 \quad \text{if} \quad p > 2,
\end{aligned}
\tag{6.2}
$$

where $\alpha_{kr} = \nu^r (-1/\lambda)^{k-r} {}_{(k-1)}C_{(r-1)}$ and ${}_pC_q$ is the binomial coefficient, then $(K_0 - \lambda^{-1} I) z_k = z_{k-1}$ for $k = 0, \dots, (p-1)$, where $z_{-1} = 0$. It follows that $(K_0 - \lambda^{-1} I)^p u = 0$, and hence we conclude that $M_\lambda \subset \mathcal{G}_{1/\lambda}$.

Conversely, suppose that λ is a characteristic value of K_0. Then it is a simple matter to verify that λ is a non-zero eigenvalue of the problem (2.1) and $N(I - \lambda K_0) \subset N_\lambda \subset M_\lambda$. Hence if λ^{-1} is a semi-simple eigenvalue of K_0, then $\mathcal{G}_{1/\lambda} \subset M_\lambda$. Now suppose that λ^{-1}

is not a semi-simple eigenvalue of K_0, that $z \in \mathcal{G}_{1/\lambda}$ and that $z \notin N(I - \lambda K_0)$. Then there exists the integer $p > 1$ such that $(K_0 - \lambda^{-1}I)^p z = 0$ and $(K_0 - \lambda^{-1}I)^{p-1}z \neq 0$. Let us put $(K_0 - \lambda^{-1}I)z_k = z_{k-1}$ for $k = 0, \dots, (p-1)$, where $z_{p-1} = z, z_{-1} = 0$, and define the vectors $\{u_k\}_0^{p-1}$ according to (6.2). Then it is not difficult to verify that $(A - \lambda T)u_k = Tu_{k-1}$ for $k = 0, \dots, (p-1)$, where $u_{-1} = 0$, and hence it follows that $z \in M_\lambda$. Thus we conclude that $\mathcal{G}_{1/\lambda} \subset M_\lambda$.

In light of the foregoing results, we are now in a position to exploit the fact that K_0 is a compact selfadjoint operator in the Pontrjagin space V_0 to prove some of the assertions of the theorem. Indeed, it follows immediately (see Theorems 3.4-5) that the eigenvalues of (2.1) are each of finite algebraic multiplicity and form an at most denumerably infinite subset of $\mathbb{C}$ having no finite points of accumulation, while the assertions concerning the orthogonality of the principal subspaces, the symmetry of the eigenvalues with respect to the real axis, and the relationship between M_λ and $M_{\bar{\lambda}}$ follow from [**47**, pp.35, 133-134], Theorems 3.4-5, Theorem 3.8, and (4.7). Moreover, if there are non-zero eigenvalues of the problem (2.1) whose corresponding eigenspaces are not definite with respect to the inner product $(\; , \;)_T$, and hence in light of (4.7), not definite with respect to the inner product $\langle \; , \; \rangle$, then it follows from [**47**, Lemma 2.1, p.4] that there exists the sequence $\{\lambda_j\}$ of distinct eigenvalues of (2.1), consisting of all non-real eigenvalues λ_j, if any, for which $Im\lambda_j > 0$ and all real eigenvalues $\lambda_j \neq 0$ for which N_{λ_j} is not definite, if such eigenvalues exist, as well as the sequence $\{u_j\}$ of corresponding eigenvectors which satisfy $\langle u_j, u_k \rangle = 0$ for all pairs j, k. Since the u_j are linearly independent and span a neutral subspace of the Pontrjagin space V_0, it follows from [**47**, Lemma 1.1, p.184] that the number of such eigenvalues λ_j cannot exceed κ. Thus in light of the foregoing results and those of Chapter 3, we conclude that there are at most $2\kappa + 1$ distinct eigenvalues of the problem (2.1) whose corresponding eigenspaces are not definite with respect to the inner product $(\; , \;)_T$, and hence (see [**47**, Lemma 3.8, p.35]) not real and semi-simple.

In view of Theorems 1.1, 2.5-6, 3.4-5, and the foregoing results, the proof of the theorem is complete.

Remark 6.3. We have seen in the proof of Theorem 6.1 that the non-zero eigenvalues of the problem (2.1) are precisely the characteristic values of K_0 and if λ is such an eigenvalue, then $N_\lambda = N(I - \lambda K_0)$. Hence it follows from (4.7) that N_λ is definite with respect to the inner product $(\; , \;)_T$ if and only if it is definite with respect to the inner product $\langle \; , \; \rangle$ (note from [**47**, Lemma 3.8, p.35] that if N_λ is definite in either sense, then λ must be real and semi-simple). Moreover, if N_λ is definite, then N_λ is positive (resp. negative) definite with respect to the inner product $\langle \; , \; \rangle$ if and only if $sgn(u, u)_T = sgn\lambda$

(resp. $sgn(u,u)_T = -sgn\lambda$) for every $u \neq 0$ in N_λ, and we can always choose a basis $\{u_j\}_1^p$ of $M_\lambda = N_\lambda$ satisfying $(u_j, u_k)_T = \delta_{jk} sgn\lambda$ (resp. $(u_j, u_k)_T = -\delta_{jk} sgn\lambda$), where δ_{jk} denotes the Kronecker delta.

Theorem 6.2. The number of non-zero, non-semi-simple eigenvalues of the problem (2.1) does not exceed κ. Moreover, the number of non-real eigenvalues, where each such eigenvalue is repeated according to its algebraic multiplicity, does not exceed 2κ.

Proof. Since K_0 is a selfadjoint operator in the Pontrjagin space V_0, the assertions of the theorem are certainly true for the eigenvalues of K_0 [**47**, Theorems 4.6 and 4.8, p.191]. On the other hand, we have shown in the proof of Theorem 6.1 that λ is a non-zero eigenvalue of the problem (2.1) if and only if λ is a characteristic value of K_0 and that M_λ is the principal subspace of K_0 corresponding to the eigenvalue $1/\lambda$. The theorem now follows from these results.

We have seen that if λ is a non-real eigenvalue of the problem (2.1), then so is $\overline{\lambda}$ and M_λ and $M_{\overline{\lambda}}$ are finite dimensional subspaces of the Pontrjagin space (resp. of the indefinite inner product space) V_0 which form a dual pair.

Theorem 6.3. If λ is a non-real eigenvalue of the problem (2.1), then $\dim M_\lambda = \dim M_{\overline{\lambda}}$. Moreover, $M_{\overline{\lambda}}$ (resp. $N_{\overline{\lambda}}$) is composed of those vectors in $\mathcal{H}$ which are precisely the complex conjugates of the vectors of M_λ (resp. N_λ).

Proof. The first assertion of the theorem follows from [**47**, Lemma 10.3, p.21]. Before proving the remaining assertions, let us observe that if $u \in H^p(\Omega)$ for some $p \in \mathbb{N}$, $\phi \in C_0^\infty(\Omega), \alpha = (\alpha_1, \alpha_2)$ is a multi-index satisfying $0 \leq |\alpha| = \alpha_1 + \alpha_2 \leq p$, and $D^\alpha = D_1^{\alpha_1} D_2^{\alpha_2}$, then $(\overline{u}, D^\alpha \phi) = (-1)^{|\alpha|}(\overline{D^\alpha u}, \phi)$, where $\overline{u}(x)$ denotes the complex conjugate of $u(x)$. In other words, a distributional derivative of $\overline{u}$ of order $\leq p$ is precisely the complex conjugate of the same distributional derivative of u. It follows immediately from Definition 2.1 and Theorem 2.4 (note that the boundary operators in (2.1b) are real differential operators) that if $u \in V$ (resp. if $u \in D(A)$), then so does $\overline{u}$. Furthermore, it also follows from Theorem 3.4 and the proof of Theorem 3.5 that if $u \in V_0$ (resp. if $u \in D(A_0)$), then so does $\overline{u}$.

Let us now prove the final assertions of the theorem. Accordingly, if $u \in N_\lambda$, then $(A - \lambda T)u = 0$, and hence it follows from Theorem 2.4 that $(Lu)(x) - \lambda\omega(x)u(x) = 0$ almost everywhere in Ω. This shows that $\overline{(Lu)(x)} - \overline{\lambda}\omega(x)\overline{u(x)} = 0$ almost everywhere in Ω, and hence, in light of what has been said in the previous paragraph and of the fact that L is a real differential operator, we conclude that $(L\overline{u})(x) - \overline{\lambda}\omega(x)\overline{u}(x) = 0$ almost everywhere in Ω. This shows that $(A - \overline{\lambda}T)\overline{u} = 0$ and so $\overline{u} \in N_{\overline{\lambda}}$. Conversely, if $u \in N_{\overline{\lambda}}$, then similar arguments show that $\overline{u} \in N_\lambda$, and hence the assertion of the theorem concerning N_λ and $N_{\overline{\lambda}}$ now follows.

Suppose next that λ is not a semi-simple eigenvalue of the problem (2.1) and that $u \in M_\lambda \backslash N_\lambda$. Then there exist vectors $\{u_k\}_0^{p-1}$ in M_λ, where $p > 1, u_{p-1} = u$, and $u_0 \neq 0$, such that $(A - \lambda T)u_k = Tu_{k-1}$ for $k = 0, \ldots, (p-1)$, where $u_{-1} = 0$. Arguing as we did in the previous paragraph, we can easily establish that $(A - \overline{\lambda}T)\overline{u}_k = T\overline{u}_{k-1}$ for $k = 0, \ldots, (p-1)$, and hence it follows that $\overline{u} \in M_{\overline{\lambda}}$. This last result, together with the fact that $\dim M_\lambda = \dim M_{\overline{\lambda}}$, proves the assertion of the theorem concerning M_λ and $M_{\overline{\lambda}}$.

Lemma 6.1. If λ is a non-real eigenvalue of the problem (2.1) and $u \in M_\lambda$, then $(A - \overline{\lambda}T)\overline{u} = \overline{(A - \lambda T)u}$ and $(K_0 - \overline{\lambda}^{-1}I)\overline{u} = \overline{(K_0 - \lambda^{-1}I)u}$.

Proof. We may argue as in the proof of Theorem 6.3 to show that if $(A - \lambda T)u = f$, then $(A - \overline{\lambda}T)\overline{u} = \overline{f}$, and this proves the first assertion of the lemma. Turning to the second assertion, we have seen in the proof of Theorem 6.1 that λ^{-1} is an eigenvalue of K_0 and that M_λ is precisely the principal subspace of K_0 corresponding to λ^{-1}. Therefore if we put $(K_0 - \lambda^{-1}I)u = v$, then $v \in M_\lambda$. This shows that $(A - \lambda T)u = -\lambda Av$, and hence we may now argue as we did in the proof of Theorem 6.3 to show that $(A - \overline{\lambda}T)\overline{u} = -\overline{\lambda}A\overline{v}$. Since we know from Theorem 6.3 that $\overline{u}, \overline{v} \in M_{\overline{\lambda}}$, it follows from this last equation that $(K_0 - \overline{\lambda}^{-1}I)\overline{u} = \overline{v}$, which is what we wanted to prove.

In the following theorem we again consider the principal subspaces of the problem (2.1) as subspaces of the indefinite inner product space V.

Theorem 6.4. Suppose that λ is a non-real eigenvalue of the problem (2.1). Then λ (resp. $\overline{\lambda}$) is semi-simple if and only if N_λ and $N_{\overline{\lambda}}$ form a dual pair. Moreover, whether λ (resp. $\overline{\lambda}$) is semi-simple or not, $M_\lambda \dot{+} M_{\overline{\lambda}}$ may be decomposed into the orthogonal direct sum of subspaces $\{\mathcal{L}_j\}_1^{n_\lambda}$, where each $\mathcal{L}_j$ is the direct sum of two subspaces $\mathcal{L}_j^+ \subset M_\lambda$ and $\mathcal{L}_j^- \subset M_{\overline{\lambda}}$ having bases $\{y_{jk}\}$ and $\{\overline{y}_{jk}\}$, $k = 0, \ldots, (p(j)-1)$, respectively, which satisfy: (1) $(A - \lambda T)y_{jk} = Ty_{j,k-1}$, $(A - \overline{\lambda}T)\overline{y}_{j,k} = T\overline{y}_{j,k-1} (y_{j,-1} = \overline{y}_{j,-1} = 0)$, (2) $(y_{jk}, \overline{y}_{jr})_T = 1$ or 0 according to whether $k + r = p(j) - 1$ or $k + r \neq p(j) - 1$,

(3) $y_{jk}, \overline{y}_{jk} \in C^1(I^2)$ and satisfies the boundary conditions (1.2) and (1.4), and where $p(1) \geq p(2) \geq \cdots \geq p(n_\lambda) \geq 1$ and $\sum_{j=1}^{n_\lambda} p(j) \leq \kappa$.

Remark 6.4. In view of Theorem 6.1, it is clear that the above theorem implies that M_λ (resp. $M_{\overline{\lambda}}$) is the orthogonal direct sum of the subspaces $\{\mathcal{L}_j^+\}_1^{n_\lambda}$ (resp. $\{\mathcal{L}_j^-\}_1^{n_\lambda}$).

Proof of Theorem 6.4. Let $\mu = \lambda^{-1}$. Then we know from the proof of Theorem 6.1 that there exists a $p \in \mathbb{N}$ such that $(K_0 - \mu I)^p M_\lambda = 0$ and $(K_0 - \mu I)^{p-1} M_\lambda \neq 0$. Note also from Theorem 6.3 and Lemma 6.1 that $(K_0 - \overline{\mu} I)^p M_{\overline{\lambda}} = 0$ and $(K_0 - \overline{\mu} I)^{p-1} M_{\overline{\lambda}} \neq 0$. If in M_λ we now introduce the form $[y, z] = \big((K_0 - \mu I)^{p-1} y, \overline{z}\big)_T$, then it follows from Lemma 6.1 that $[y, z]$ is a bilinear form and that $[y, z] = [z, y]$. Moreover, we assert that there exists a $z \in M_\lambda$ such that $[z, z] \neq 0$. Indeed, if $[z, z] = 0$ for every $z \in M_\lambda$, then it follows from the equation

$$[y + tz, y + tz] = [y, y] + 2t[y, z] + t^2[z, z], \; y, z \in M_\lambda, \; t \in \mathbb{C},$$

that $[y, z] = 0$ for every $y, z \in M_\lambda$. But by hypothesis there exists a $z_0 \in M_\lambda$ such that $(K - \mu I)^{p-1} z_0 \neq 0$, and hence, in light of Theorems 6.1 and 6.3 we arrive at the contradiction that $\big((K - \mu I)^{p-1} z_0, y\big)_T = 0$ for every $y \in M_{\overline{\lambda}}$. This proves the assertion, and clearly we can normalize z so that $[z, z] = 1$.

Let $(K_0 - \mu I) z_k = z_{k-1}$ for $k = 0, \ldots, (p-1)$, where $z_{p-1} = z$ and $z_{-1} = 0$. Then it follows from Lemma 6.1 that $(z_j, \overline{z}_k)_T$ equals 0 if $p > 1$ and $j + k < p - 1$ and equals 1 if $j + k = p - 1$. Hence if we define the vectors $\{u_k\}_0^{p-1}$ according to (6.2), then we know from the proof of Theorem 6.1 that $(A - \lambda T) u_k = T u_{k-1}$ for $k = 0, \ldots, (p-1)$, where $u_{-1} = 0$, while it is also clear that $(u_j, \overline{u}_k)_T$ equals zero if $p > 1$ and $j + k < p - 1$ and equals ν^{p-1} if $j + k = p - 1$, where $\nu = -\lambda^{-2}$. Observe also from Lemma 6.1 that if $p > 1$ and $j < p - 1$, then $(u_j, \overline{u}_k)_T = (u_{j+1}, \overline{u}_{k-1})_T$. Hence if $\{\delta_k\}_0^{p-1}$ are complex constants and

$$y_k = \sum_{j=0}^{k} \delta_{p-1-k+j} u_j \quad \text{for } k = 0, \ldots, (p-1), \tag{6.3}$$

then it follows that

$$\begin{aligned} (y_j, \overline{y}_k)_T &= 0 \quad \text{if } p > 1 \text{ and } j + k < p - 1 \\ &= \delta_{p-1}^2 \nu^{p-1} \quad \text{if } j + k = p - 1 \end{aligned} \tag{6.4}$$

$$= \left[2\delta_{p-1}\delta_{p-1-\ell} + \sum_{i=1}^{\ell-1} \delta_{p-1-i}\delta_{p-1-\ell+i}\right]\nu^{p-1} +$$

$$+ \sum_{r=1}^{\ell}(u_{p-1}, \overline{u}_r)_T \sum_{i=r}^{\ell} \delta_{p-1+r-i}\delta_{p-1-\ell+i} \tag{6.5}$$

$$\text{if } p > 1, j + k = p - 1 + \ell, \text{ and } 1 \leq \ell \leq p - 1,$$

where the summation in the first term on the right side of (6.5) does not appear if $\ell = 1$. Let $\delta_{p-1} = 1/\nu^{(p-1)/2}$, where the principal determination of the square root is taken. Then it follows from (6.4) that $(y_j, \overline{y}_k)_T = 1$ if $j + k = p - 1$, while (6.5) shows that if $p > 1$, then we can choose the $\delta_{p-1-\ell}, \ell = 1, \ldots, (p-1)$, so that $(y_j, \overline{y}_k)_T = 0$ if $j + k > p - 1$. Choosing the $\delta_k, k = 0, \ldots, (p-2)$, in this way, we therefore conclude that $(y_j, \overline{y}_k)_T$ equals 1 or 0 according to whether $j + k = p - 1$ or $j + k \neq p - 1$. Observe also from (6.3) that $(A - \lambda T)y_k = Ty_{k-1}$ for $k = 0, \ldots, (p-1)$, where $y_{-1} = 0$.

Let $\mathcal{L}_1^+ = span\{y_k\}_0^{p-1}, \mathcal{L}_1^- = span\{\overline{y}_k\}_0^{p-1}, \mathcal{L}_1 = \mathcal{L}_1^+ \dot{+} \mathcal{L}_1^-, p(1) = p$, and put $y_{1k} = y_k, \overline{y}_{1k} = \overline{y}_k$ for $k = 0, \ldots, (p-1)$. Observe from what has just been shown and Theorem 6.1 that $\mathcal{L}_1$ is a non-degenerate subspace of V, and moreover, if $\mathcal{L}_1 = M_\lambda \dot{+} M_{\overline{\lambda}}$, then one part of the theorem is proved. If $\mathcal{L}_1 \neq M_\lambda \dot{+} M_{\overline{\lambda}}$, then let $\mathcal{N}_1$ denote the orthogonal companion of $\mathcal{L}_1$ in V and put $\mathcal{L}^+ = \mathcal{N}_1 \cap M_\lambda, \mathcal{L}^- = \mathcal{N}_1 \cap M_{\overline{\lambda}}$. Then it follows from [**47**, Corollary 11.9, p.26] that $M_\lambda \dot{+} M_{\overline{\lambda}}$ is the orthogonal direct sum of $\mathcal{L}_1$ and $\mathcal{L}^+ \dot{+} \mathcal{L}^-$, while Theorem 6.1 and (4.7) show that $\mathcal{L}^+$ and $\mathcal{L}^-$ form a dual pair and each is invariant under K_0. Hence we can argue with $\mathcal{L}^+ \dot{+} \mathcal{L}^-$ as we argued with $M_\lambda \dot{+} M_{\overline{\lambda}}$ above, and then continue with these arguments, if necessary, to arrive at the decompositions of $M_\lambda \dot{+} M_{\overline{\lambda}}$ and the $\mathcal{L}_j$ as asserted in the theorem as well as verifying that the y_{jk} and $\overline{y}_{jk}$ do indeed satisfy the conditions (1) and (2) stated there. The assertion that $p(1) \geq p(2) \geq \cdots \geq p(n_\lambda) \geq 1$ follows from the construction, while the assertion that $\sum_{j=1}^{n_\lambda} p(j) \leq \kappa$ follows from [**47**, Lemma 1.1, p.184] and the fact that M_λ is a neutral subspace of the Pontrjagin space V_0 (see the proof of Theorem 6.1). To prove that the y_{jk} and $\overline{y}_{j,k}$ satisfy condition (3) of the theorem, let us fix our attention upon a particular pair of integers j, k, $1 \leq j \leq n_\lambda$, $0 \leq k < p(j)$, and observe that since $(A - \lambda T)y_{jk} = Ty_{j,k-1}$, we have $B(y_{jk}, v) = (f, v)$ for every $v \in V$, where $f = \lambda T y_{jk} + T y_{j,k-1}$. Furthermore, since we know from the Sobolev imbedding theorem [**1**, Theorem 5.4, p.97] that $f \in L^\infty(\Omega)$, it follows from Theorem 2.3 that $y_{jk} \in W^{2,r}(\Omega)$ for every r satisfying $0 < r < \infty$. Hence we conclude from the Sobolev imbedding theorem that we may modify y_{jk} on a set of measure zero and extend y_{jk} to all of I^2 by continuity so that $y_{jk} \in C^1(I^2)$; and since $y_{jk} \in D(A)$, it also follows from Theorem 2.4 that y_{jk} satisfies the boundary conditions (1.2) and (1.4).

Thus all the assertions of the theorem, except the first, have now been proved. To prove the first assertion, let us suppose initially that λ is a semi-simple eigenvalue of the problem (2.1) (note from Theorem 6.3 that λ is semi-simple if and only if $\overline{\lambda}$ is). Then $M_\lambda = N_\lambda$, $M_{\overline{\lambda}} = N_{\overline{\lambda}}$, and hence it follows from Theorem 6.1 that N_λ and $N_{\overline{\lambda}}$ form a dual pair of subspaces of V. Conversely, if N_λ and $N_{\overline{\lambda}}$ form a dual pair of subspaces of V, then it follows from the results already proved that $p(j) = 1$ for $j = 1, \ldots, n_\lambda$, and hence $M_\lambda = N_\lambda$. This completes the proof of the theorem.

If we again consider the principal subspaces of the problem (2.1) as subspaces of the indefinite inner product space V, then arguments similar to those used in the proof of Theorem 6.4 together with Theorem 4.9 of [**47**, p.191] show that

Theorem 6.5. Suppose that $\lambda \neq 0$ is a real eigenvalue of the problem (2.1) such that N_λ is not a definite subspace of V. Then λ is semi-simple if and only if the inner product $(\, , \,)_T$ is not degenerate on N_λ. Moreover, whether λ is semi-simple or not, M_λ may be decomposed into the orthogonal direct sum of subspaces $\{\mathcal{L}_j\}_1^{n_\lambda}$, where each $\mathcal{L}_j$ has a basis $\{y_{jk}\}$, $k = 0, \ldots, (p(j) - 1)$, which satisfies: (1) $(A - \lambda T)y_{jk} = Ty_{j,k-1}(y_{j,-1} = 0)$, (2) $(y_{jk}, y_{jr})_T = \rho_j (= \pm 1)$ or 0 according to whether $k + r = p(j) - 1$ or $k + r \neq p(j) - 1$, (3) $y_{jk} \in C^1(I^2)$ and satisfies the boundary conditions (1.2) and (1.4), and where $2\kappa + 1 \geq p(1) \geq p(2) \geq \cdots \geq p(n_\lambda) \geq 1$.

We have seen in the proof of Theorem 6.1 that the non-zero eigenvalues of the problem (2.1) are precisely the characteristic values of K_0 and for each such eigenvalue λ, M_λ is the principal subspace of K_0 corresponding to $1/\lambda$. Thus in particular, we see that if λ is a non-zero eigenvalue of (2.1), then $M_\lambda \subset V_0$, M_λ is invariant under K_0, and $0 \in \rho(K_0|M_\lambda)$ (recall that V_0 is a Hilbert space with respect to the strong topology and K_0 is a compact operator acting in this space). Let us now consider the principal subspaces of the problem (2.1) corresponding to its non-zero eigenvalues as subspaces of the Pontrjagin space V_0. Then it follows from Theorem 6.1 and (4.7) that if λ and μ are non-zero eigenvalues of (2.1) such that $\lambda \neq \overline{\mu}$, then M_λ and M_μ are orthogonal, while if λ is a non-real eigenvalue of (2.1), then M_λ is neutral. Note also from Remark 6.3 that if $\lambda \neq 0$ is an eigenvalue of (2.1) such that N_λ is a negative definite subspace of V_0, then $sgn(u,u)_T = -sgn\lambda$ for every $u \neq 0$ in N_λ. Furthermore, if we turn our attention to Theorem 6.5, then it is a simple matter of verify that for each j, $\mathcal{L}_j$ is invariant under K_0 and $0 \in \rho(K_0|\mathcal{L}_j)$, and hence it follows from (4.7) that the $\mathcal{L}_j$ are also pairwise orthogonal subspaces of V_0. Still keeping our attention fixed upon Theorem 6.5, suppose firstly that $p(1) > 1$. Then for this case we let $j^\star$ denote the largest value of j, $1 \leq j \leq n_\lambda$, for which $p(j) > 1$ and let M_λ^0 denote the span of the subspaces $\{\mathcal{L}_j\}_1^{j\star}$. If $p(1) = 1$, then put $M_\lambda^0 = 0$. If $p(n_\lambda) = 1$, then let M_λ^- denote the span of those $\mathcal{L}_j$ for which $p(j) = 1$ and $(y_{j0}, y_{j0})_T = -sgn\lambda$ (or equivalently, $\langle y_{j0}, y_{j0}\rangle < 0$, in view of (4.7)) if such $\mathcal{L}_j$ exist, and put $M_\lambda^- = 0$ otherwise. If $p(n_\lambda) > 1$, then put $M_\lambda^- = 0$. Then M_λ^0 and M_λ^- are pairwise orthogonal subspaces of V_0 and M_λ^- is negative definite.

Theorem 6.6. $[\sum_\lambda(\dim M_\lambda^0 + \dim M_\lambda^-) + \sum_\lambda \dim M_\lambda] \leq 3\kappa$, where the first summation is over all λ satisfying the hypotheses of Theorem 6.5 and the second summation is over all non-zero eigenvalues λ of the problem (2.1) for which either λ is not real or λ is real and N_λ is a negative definite subspace of the Pontrjagin space V_0, and where the first or second summation is to be omitted if eigenvalues of the kinds asserted do not exist.

Proof. Returning to the discussion preceding the theorem, let us assume that $M_\lambda^0 \neq 0$ and let

$$k(j) = (p(j) - 2)/2 \text{ if } p(j) \text{ is even}$$

$$= (p(j) - 3)/2 \text{ if } p(j) \text{ is odd}$$

for $j = 1, \ldots, j^\star$. Observe from (4.7) and Theorem 6.5 that if $1 \leq j \leq j^\star$ and $0 \leq k$, $r \leq k(j)$, then $\langle y_{jk}, y_{jr}\rangle = (\lambda y_{jk} + y_{j,k-1}, y_{jr})_T = 0$. Thus if $M_\lambda^1 = span\{y_{jk}\}$, $j = 1, \ldots, j^\star$, $k = 0, \ldots, k(j)$, then it follows that M_λ^1 is a neutral subspace of V_0. If $M_\lambda^0 = 0$, then put $M_\lambda^1 = 0$.

Let us suppose next that the problem (2.1) has eigenvalues $\lambda \neq 0$ for which: (1) the hypotheses of Theorem 6.5 are satisfied, (2) $Im\lambda \neq 0$, and (3) λ is real and N_λ is a negative definite subspace of V_0. Then it is clear that the M_λ^1, the M_λ^-, the M_λ for which $Im\lambda > 0$, and the M_λ for which $\lambda \neq 0$ is real and N_λ is negative definite are linearly independent subspaces of V_0 and the span of these subspaces is a negative subspace of V_0. Hence it follows from [**47**, Lemma 1.1, p.184] that

$$\sum_\lambda (\dim M_\lambda^1 + \dim M_\lambda^-) + \sum_\lambda \dim M_\lambda \leq \kappa, \tag{6.6}$$

where the first summation is over all λ satisfying the hypotheses of Theorem 6.5 and the second summation is over all non-zero eigenvalues λ of (2.1) for which either $Im\lambda > 0$ or λ is real and N_λ is a negative definite subpsace of V_0. Writing $p_\lambda(j)$ for $p(j)$, $j = 1, \ldots, n_\lambda$, in Theorem 6.5, it follows from (6.6) and the definitions of the $k(j)$ that

$$\sum_\lambda \sum_{j=1}^{n_\lambda} 2^{-1}(p_\lambda(j) - 1) + \sum_\lambda \dim M_\lambda^- + \sum_\lambda \dim M_\lambda \leq \kappa, \tag{6.7}$$

and that the number of non-zero terms occurring in the double summation does not exceed κ, where in the first two expressions on the left side of (6.7), $\sum_\lambda$ denotes summation over all λ satisfying the hypotheses of Theorem 6.5, while the third expression, $\sum_\lambda \dim M_\lambda$, is the same as in (6.6). In light of Theorem 6.3, the assertion of the theorem now follows for the case under consideration here, while the proof of the theorem for the general case follows from obvious modification of the foregoing arguments.

If the problem (2.1) has non-real eigenvalues, then let $\mathcal{M}_1$ denote the span of all the M_λ for which $Im\lambda \neq 0$; otherwise let $\mathcal{M}_1 = 0$. If the problem (2.1) has real non-zero eigenvalues whose corresponding eigenspaces are not definite subspaces of the indefinite inner product space V (and hence of the Pontrjagin space V_0), then let $\mathcal{M}_2$ denote the span of the M_λ corresponding to such eigenvalues; otherwise let $\mathcal{M}_2 = 0$. If the problem (2.1) has real non-zero eigenvalues whose corresponding eigenspaces are negative definite subspaces of the Pontrjagin space V_0, then let $\mathcal{M}_3$ denote the span of the M_λ corresponding to such eigenvalues; otherwise let $\mathcal{M}_3 = 0$.

Proposition 6.3. The $\mathcal{M}_j$, $j = 1, 2, 3$, are pairwise orthogonal, linearly independent subspaces of the Pontrjagin space V_0 (resp. of the indefinite inner product space V), and if M denotes their orthogonal direct sum, then M is non-degenerate. Moreover, if $M \neq 0$, then M is either an indefinite or a negative definite subspace of the Pontrjagin space V_0, while M is also invariant under K_0 and $0 \in \rho(K_0|M)$.

Proof. If the $\mathcal{M}_j$ are not all zero, then we may appeal to the orthogonality properties of the bases constructed in Remark 6.3 and Theorems 6.4-5 as well as to the assertions concerning orthogonality made in Theorem 6.1 to show that these subspaces are linearly independent and pairwise orthogonal in the indefinite inner product space V and that M is a non-degenerate subspace of V. Furthermore, since we already know (see the remarks preceding Theorem 6.6) that if λ is a non-zero eigenvalue of the problem (2.1), then $M_\lambda \subset V_0$, M_λ is invariant under K_0, and $0 \in \rho(K_0|M_\lambda)$, it follows immediately that $\mathcal{M}_j$

(resp. M) is contained in V_0, $\mathcal{M}_j$ (resp. M) is invariant under K_0, and $0 \in \rho(K_0|\mathcal{M}_j)$ (resp. $0 \in \rho(K_0|M)$) for $j = 1, 2, 3$. Hence in light of these results and (4.7), all the assertions of the proposition, except the one concerning the definiteness or indefiniteness of M in V_0, now follow.

Suppose next that $M \neq 0$. If $\mathcal{M}_1 = \mathcal{M}_2 = 0$, then it is clear that M is a negative definite subspace of the Pontrjagin space V_0. Let us now show that if $\mathcal{M}_1 \neq 0$, then M contains positive as well as negative vectors. To this end let λ be a non-real eigenvalue of the problem (2.1), and turning our attention to Theorem 6.4, let us fix a $j, 1 \leq j \leq n_\lambda$, and put $z = c_1 y_{j,0} + c_2 \overline{y}_{j,p(j)-1}$, where the c_r are constants. Then it follows from Theorem 6.4 and the proof of Theorem 6.1 that $\langle z, z \rangle = 2Re(c_1 \overline{c}_2 \lambda)$, and hence if we take $c_2 = sgn\lambda$, then we see that $\langle z, z \rangle > 0$ if $c_1 > 0$ and $\langle z, z \rangle < 0$ if $c_1 < 0$, which is the required result. Similarly, we can show that if $\mathcal{M}_2 \neq 0$, then M contains positive and negative vectors. Finally, since M is a non-degenerate subspace of the Pontrjagin space V_0, we know from [**47**, Corollary 11.8, p.26] that M is the orthogonal direct sum of a positive definite subspace and a negative definite subspace, and we have just seen that neither of these two subspaces can reduce to 0 if $\mathcal{M}_1 \neq 0$ or $\mathcal{M}_2 \neq 0$. This completes the proof of the proposition.

Suppose that λ is a non-real eigenvalue of the problem (2.1). Then fixing our attention upon Theorem 6.4, let us henceforth write in this theorem $p_\lambda(j)$ for $p(j)$ and y_{jk}^λ (resp. $\overline{y_{jk}^\lambda}$) for y_{jk} (resp. $\overline{y}_{jk}$) to show the dependence of these expressions on λ. Also let

$$I_\lambda = \Big\{(j,k) | k = 0, \ldots, (p_\lambda(j) - 1), j = 1, \ldots, n_\lambda\Big\}, \tag{6.8}$$

and for $f \in \mathcal{H}$ and $(j,k) \in I_\lambda$, put

$$c_{jk}^\lambda(f) = \left(f, \overline{y_{j,p_\lambda(j)-1-k}^\lambda}\right)_T, \quad d_{jk}^\lambda(f) = \left(f, y_{j,p_\lambda(j)-1-k}^\lambda\right)_T. \tag{6.9}$$

Assume next that $\lambda \neq 0$ is a real eigenvalue of (2.1) such that N_λ is not a definite subspace of the indefinite inner product space V. Then fixing our attention upon Theorem 6.5, let us henceforth write in this theorem $p_\lambda(j)$ for $p(j)$, y_{jk}^λ for y_{jk}, and let I_λ be defined by (6.8). Also for $f \in \mathcal{H}$ and $(j,k) \in I_\lambda$, let

$$c_{jk}^\lambda(f) = \left(f, y_{j,p_\lambda(j)-1-k}^\lambda\right)_T \Big/ \left(y_{jk}^\lambda, y_{j,p_\lambda(j)-1-k}^\lambda\right)_T. \tag{6.10}$$

Finally, suppose that $\lambda \neq 0$ is a real eigenvalue of (2.1) such that N_λ is a negative definite subspace of the Pontrjagin space V_0. Then fixing our attention upon Remark 6.3, let us henceforth write in this remak n_λ for p, u_j^λ for u_j, and for $f \in \mathcal{H}$ and $1 \leq j \leq n_\lambda$ let

$$c_j^\lambda(f) = (f, u_j^\lambda)_T / (u_j^\lambda, u_j^\lambda)_T. \tag{6.11}$$

Notation. We let $\mathcal{H}_1 = (TM)^\perp \cap \mathcal{H}_0$ (see the notation preceding Theorem 3.4) and let $\langle\dot{+}\rangle$ (resp. $(\dot{+})$) denote orthogonal direct sum of subspaces of the Pontrjagin space V_0 (resp. of the indefinite inner product space V).

Proposition 6.4. It is the case that: (1) $\mathcal{H}_0 = M \dot{+} \mathcal{H}_1$ and $\mathcal{H}_1$ is a closed subspace of the Hilbert space $\mathcal{H}_0$ and $N(T) \subset \mathcal{H}_1$, (2) $V_0 = M\langle\dot{+}\rangle V_1$, where $V_1 = V_0 \cap \mathcal{H}_1$ is a closed, non-degenerate subspace of the Pontrjagin space V_0 which is dense in $\mathcal{H}_1$ with respect to the norm $\|\ \|$, (3) $V_0 = M(\dot{+})V_1$, and (4) K_0 is reduced by the decomposition $V_0 = M\langle\dot{+}\rangle V_1$ (resp. $V_0 = M(\dot{+})V_1$) and $N(K_0) \subset V_1$.

Proof. It is clear that we need only prove the proposition for the case $M \neq 0$. Accordingly, for $f \in \mathcal{H}_0$, let

$$\begin{aligned} F = f - \sum_\lambda \sum_{j,k \in I_\lambda} \left[c_{jk}^\lambda(f) y_{jk}^\lambda + d_{jk}^\lambda(f) \overline{y_{jk}^\lambda} \right] - \\ - \sum_\lambda \sum_{j,k \in I_\lambda} c_{jk}^\lambda(f) y_{jk}^\lambda - \sum_\lambda \sum_{j=1}^{n_\lambda} c_j^\lambda(f) u_j^\lambda, \end{aligned} \tag{6.12}$$

where in the first double sum on the right side of (6.12), $\sum_\lambda$ denotes summation over all those distinct eigenvalues λ of the problem (2.1) for which $Im\lambda > 0$, the $c_{jk}^\lambda(f)$, $d_{jk}^\lambda(f)$ are given by (6.9), and the remaining terms are defined in the statements just preceding (6.9), while in the second double sum, $\sum_\lambda$ denotes summation over those distinct eigenvalues λ of (2.1) for which λ is real, non-zero, and N_λ is not a definite subspace of the indefinite inner product space V, the $c_{jk}^\lambda(f)$ are given by (6.10), and the remaining terms are defined in the statements just preceding (6.10), and lastly, in the third double sum, $\sum_\lambda$ denotes summation over those distinct eigenvalues λ of (2.1) for which λ is real, non-zero, and N_λ is a negative subspace of the Pontrjagin space V_0, the $c_j^\lambda(f)$ are given by (6.11), and the remaining terms are defined in the statements just preceding (6.11) (note that a double summation is to be omitted if eigenvalues of the kind asserted do not exist). It follows immediately from Theorems 6.1, 6.4-5, and Remark 6.3 that $F \in \mathcal{H}_1$. Thus we see that $\mathcal{H}_0$ is the vector sum of M and $\mathcal{H}_1$, and hence, in light of Proposition 6.3, we conclude that $\mathcal{H}_0$ is the direct sum of these two subspaces. That $\mathcal{H}_1$ is a closed subspace of $\mathcal{H}_0$ follows from the fact that $(TM)^\perp$ and $\mathcal{H}_0$ are closed subspaces of $\mathcal{H}$. If $u \in N(T)$, then we know from Theorem 3.6 that $u \in \mathcal{H}_0$, and consequently $u = u_1 + u_2$, where $u_1 \in M$ and $u_2 \in \mathcal{H}_1$. Since $Tu_1 = -Tu_2$, we conclude that $u_1 \in M \cap \mathcal{H}_1$, and hence $u_1 = 0$.

The foregoing arguments show that V_0 is the direct sum of M and V_1, while it follows from (4.7) and Proposition 6.3 that M and V_1 are orthogonal subspaces of V_0. Thus $V_0 = M\langle\dot{+}\rangle V_1$, and hence we conclude from [**47**, Theorem 3.5, p.104] that V_1 is closed and non-degenerate. To show that V_1 is a dense subspace of the Hilbert space $\mathcal{H}_1$, let P denote the projection on $\mathcal{H}_0$ mapping $\mathcal{H}_0$ onto $\mathcal{H}_1$ along M. Then if $u \in \mathcal{H}_1$ and ϵ is any positive number, we know from Theorem 3.6 that there is a $v \in V_0$ such that $\|u - v\| < \epsilon$. Hence $Pv \in V_1$ and $\|u - Pv\| \leq \epsilon\|P\|$, where in the right side of this inequality, $\| \ \|$ denote the norm in $\mathcal{L}(\mathcal{H}_0)$.

It follows from the foregoing results, (4.7), and Proposition 6.3 that $V_0 = M(\dot{+})V_1$. Now suppose that $u \in M$ and $v \in V_1$. Then in light of Proposition 6.3 and the results already proved, we have $\langle u, K_0 v\rangle = \langle K_0 u, v\rangle = 0$, and so we conclude that V_1 is invariant under K_0. Thus it follows from Proposition 6.3 that K_0 is reduced by the decomposition $V_0 = M\langle\dot{+}\rangle V_1$, and hence by the decomposition $V_0 = M(\dot{+})V_1$. Finally, if $u \in N(K_0)$, then $u \in V_0 \cap N(T) \subset V_0 \cap \mathcal{H}_1 = V_1$, and this completes the proof of the proposition.

Remark 6.5. Propositions 6.2 and 6.4 assure us that when V_1, considered only as a vector space, is equipped with the inner product $(\ , \ )_{1,\Omega}$, then it becomes a Hilbert space.

Proposition 6.5. V_1, considered as a subspace of the Pontrjagin space V_0, is also a Pontrjagin space whose index κ_1 satisfies $0 \leq \kappa_1 \leq \kappa$, and κ_1 is precisely the rank of negativity of V_1.

Proof. Clearly we need only consider the case $M \neq 0$. Then all but the last of the assertions of the proposition follow from Proposition 6.4 and [**47**, Corollary 2.3, p.186]. It also follows from Proposition 6.3 and [**47**] that M is a Pontrjagin space. Hence if $V_1 = \mathcal{L}^+\langle\dot{+}\rangle\mathcal{L}^-$ and $M = \mathcal{M}^+\langle\dot{+}\rangle\mathcal{M}^-$ are fundamental decompositions of V_1 and M, respectively, where $\mathcal{L}^+$ and $\mathcal{M}^+$ are positive definite while $\mathcal{L}^-$ and $\mathcal{M}^-$ are negative definite, then $V_0 = (\mathcal{M}^+\langle\dot{+}\rangle\mathcal{L}^+)\langle\dot{+}\rangle(\mathcal{M}^-\langle\dot{+}\rangle\mathcal{L}^-)$. This shows that if $\mathcal{L}^+$ is finite dimensional, then the rank of positivity of V_0 is finite, which contradicts the assertions made in the proof of Proposition 6.1. Thus $\mathcal{L}^+$ must be infinite dimensional, and hence κ_1 is precisely the rank of negativity of V_1, which completes the proof of the proposition.

Remark 6.6. We know from Proposition 6.3 and the proof of Proposition 6.5 that M is a Pontrjagin space whose rank of negativity is precisely $\kappa - \kappa_1$ and that K_0 is a compact selfadjoint operator acting in this space. Hence it follows from an inspection of the proofs of the theorems concerned that the assertions of Theorems 6.2 and 6.6 remain perfectly valid if κ there is replaced by $\kappa - \kappa_1$.

Proposition 6.6. The strong topology of the Pontrjagin space V_1 coincides with the Hilbert topology of V_1 induced by the norm $\| \ \|_{1,\Omega}$.

Proof. In view of Propositions 6.1, 6.4-5, it follows from [**47**, Theorem 8.5, p.96] and the remarks following Theorem 1.4 of [**47**, p.102] that the strong topology of V_1 coincides with the topology induced in V_1 by the strong topology of V_0. Hence the assertion of the proposition now follows from Proposition 6.2.

Remark 6.7. We will henceforth let $K_1 = K_0|V_1$. Then in light of Propositions 6.2, 6.4-6 and the remarks made at the beginning of this section, we see that K_1 is a compact selfadjoint in the Pontrjagin space V_1. Moreover, it follows from the proof of Theorem 6.1 and the definition of M that the eigenvalues of K_1 are all real and denumerably infinite in number, while each non-zero eigenvalue is semi-simple and its corresponding eigenspace is positive definite.

We are now in a position to establish some further results concerning the spectral properties of the problem (2.1). Accordingly, recalling the notation introduced in the statements preceding Theorem 4.2 and bearing in mind Remark 4.2, we have

Theorem 6.7. Suppose that $N(T) \cap V$ is not degenerate with respect to the inner product $B(\ ,\ )$. Then the problem (2.1) has infinitely many positive and infinitely many negative eigenvalues. Moreover, the eigenvectors of (2.1) together with their associated vectors, if any, are complete in $L^2(\Omega^\#)$ and in $L^2(\Omega^\#; |\omega(x)|dx)$.

Remark 6.8. Let $\mathcal{L} = N(T) \cap V$. Then the hypothesis of the theorem is certainly satisfied if $\mathcal{L} = 0$, and this is always the case if $|I_\omega^2| = 0$. Furthermore, it is clear from Theorem 3.6 and Proposition 6.4 that $\mathcal{L} = N(T) \cap V_1 = N(K_1)$, and hence it follows that the hypothesis of the theorem is equivalent to the requirement that $N(K_1)$ be a non-degenerate subspace of the Pontrjagin space V_1. It follows immediately that the hypothesis of the theorem is satisfied when $\kappa_1 = 0$, since in this case V_1 is a Hilbert space with respect to the inner product $\langle\ ,\ \rangle$. Note also that it is in this latter formulation that we recognize our hypothesis as being the sufficient condition given in [**39**] and [**110**] for the completeness of the root vectors of K_0 in the Pontrjagin space V_0.

Proof of Theorem 6.7. We already know that V_1 is a Hilbert space with respect to the inner product $\langle\ ,\ \rangle$ if $\kappa_1 = 0$. Let us now show that if $\kappa_1 > 0$, then the Pontrjagin space V_1 admits a fundamental decomposition $V_1 = \mathcal{L}^+ \langle\dot{+}\rangle \mathcal{L}^-$ with positive definite component $\mathcal{L}^+$ and negative definite component $\mathcal{L}^-$ such that $\mathcal{L}^- \subset N(K_1)$ and $\mathcal{L}^+$ is invariant under K_1. Accordingly, suppose that $\kappa_1 > 0$ and assume firstly that $\mathcal{L} \neq 0$, where $\mathcal{L}$ is defined in Remark 6.8. Then it follows from [**47**, Corollary 2.3, p.186] that

$$\mathcal{L} = \mathcal{L}_+ \langle\dot{+}\rangle \mathcal{L}_-, \tag{6.13}$$

where $\mathcal{L}_+$ is positive definite, $\mathcal{L}_-$ is negative definite, and $0 \leq \dim \mathcal{L}_- \leq \kappa_1$ [**47**, Lemma 1.1, p.184]. Next let $\mathcal{L}^\# = \mathcal{L}_-$ if $\mathcal{L} \neq 0$ and $\mathcal{L}^\# = 0$ otherwise. Then we know from [**47**, Theorem 2.1, p.165 and Theorem 3.2, p.169] that K_1 has an invariant maximal negative subspace $\mathcal{L}^-$ containing $\mathcal{L}^\#$, while from [**47**, Lemma 1.2, p.184] we also know that $\dim \mathcal{L}^- = \kappa_1$. Since $\sigma(K_1|\mathcal{L}^-)$ consists precisely of the point $\lambda = 0$, we conclude that $\mathcal{L} \neq 0$ and, in view of (6.13), that $\mathcal{L}_- \neq 0$. Now suppose that $\mathcal{L}^-$ is a degenerate subspace of V_1. Then there is a $u \neq 0$ in $\mathcal{L} \cap \mathcal{L}^-$ such that $\langle u, u\rangle = 0$, and hence it follows from the Schwarz inequality [**47**, Lemma 2.2, p.5] that $\langle u, v\rangle = 0$ for every $v \in \mathcal{L}^-$. But then (6.13) implies that $u = 0$, which is a contradiction. Thus we conclude that $\lambda = 0$ is a semi-simple eigenvalue of K_1, $\mathcal{L}^- \subset N(K_1)$, and $\mathcal{L}^-$ is negative definite. If we now let $\mathcal{L}^+$ denote the orthogonal companion of $\mathcal{L}^-$ in the Pontrjagin space V_1, then it follows from [**47**, Lemma 2.1, p.186 and Theorem 7.1, p.112] and the fact that K_1 is selfadjoint, that $\mathcal{L}^+ \langle\dot{+}\rangle \mathcal{L}^-$ is precisely the fundamental decomposition of V_1 asserted above.

Let $\mathcal{N} = V_1$ if $\kappa_1 = 0$ and $\mathcal{N} = \mathcal{L}^+$ otherwise. Then we know from above that $\mathcal{N}$ is a Hilbert space with respect to the inner product $\langle\ ,\ \rangle$ and $K_2 = K_1|\mathcal{N}$ is a compact selfadjoint operator in this space. Considering $\mathcal{N}$ as such a Hilbert space, let P^+ and P^- denote the positive and negative spectral projections, respectively, of K_2 and let $\mathcal{N}^\pm = P^\pm \mathcal{N}$. Then, since we know from the proof of Theorem 6.1 that the non-zero eigenvalues of the problem (2.1) are precisely the characteristic vectors of K_0, it is clear that in order to prove that (2.1) has infinitely many positive and negative eigenvalues we need only show that $\dim \mathcal{N}^{\pm} = \infty$. To this end let us for the remainder of this paragraph consider all spaces concerned as subspaces of the indefinite inner product space V. Then it follows from the spectral theorem and (4.7) that $\mathcal{N}$ admits a fundamental decomposition with neutral component $N(K_2)$, positive definite component $\mathcal{N}^+$, and negative definite component $\mathcal{N}^-$. Thus, in light of what has been proved above and (4.7), we see that V_1 admits a fundamental decomposition with neutral component $N(K_1)$, positive definite component $\mathcal{N}^+$, and negative definite component $\mathcal{N}^-$. Moreover, we know from Proposition 6.3 and [**47**, Corollary 11.8, p.26] that if $M \neq 0$, then M admits a fundamental decomposition, $M = \mathcal{M}^+(\dot{+})\mathcal{M}^-$, where $\mathcal{M}^+$ is positive definite and $\mathcal{M}^-$ negative definite, while Theorems 3.4-5 show that if $0 \in \sigma(A)$, then M_0 admits a fundamental decomposition, $M_0 = \mathcal{M}_0^+(\dot{+})\mathcal{M}_0^-$, where $\mathcal{M}_0^+$ is positive definite and $\mathcal{M}_0^-$ negative definite. Hence it follows from Theorems 3.4-6 and Proposition 6.4 that V admits the fundamental decomposition

$$V = N(K_1)(\dot{+})\Big(\mathcal{M}_0^+(\dot{+})\mathcal{M}^+(\dot{+})\mathcal{N}^+\Big)(\dot{+})\Big(\mathcal{M}_0^-(\dot{+})\mathcal{M}^-(\dot{+})\mathcal{N}^-\Big), \tag{6.14}$$

where $\mathcal{M}_0^{\pm}$ (resp. $\mathcal{M}^{\pm}$) is 0 if $0 \in \rho(A)$ (resp. if $M = 0$). As a consequence of (6.14) and arguments used in the proof of Proposition 3.3, it is not difficult to show that if $\mathcal{G}$ is a positive definite subspace of V, then $\mathcal{G}$ is algebraically isomorphic to a subspace of $\mathcal{M}_0^+(\dot{+})\mathcal{M}^+(\dot{+})\mathcal{N}^+$. But it is clear from the definition of ω that V contains positive definite subspaces of arbitrary large dimension, and hence we conclude that $\dim \mathcal{N}^+ = \infty$. Similarly we can show that $\dim \mathcal{N}^- = \infty$.

Returning again to the Hilbert space $\mathcal{N}$ defined above and bearing in mind (4.7), Theorem 6.1, and Remarks 6.3 and 6.7, we see that the characteristic values and characteristic vectors of K_2 may be arranged into sequences $\{\mu_j\}_1^\infty$ and $\{u_j\}_1^\infty$, respectively, where each characteristic value is repeated according to its multiplicity, $0 < |\mu_1| \leq |\mu_2| \leq \ldots, (u_j, u_k)_T = \delta_{jk} sgn\mu_j, \langle u_j, u_k\rangle = \delta_{jk}|\mu_j|$, and δ_{jk} is the Kronecker delta. Putting $v_j = |\mu_j|^{-1/2} u_j$ for $j \geq 1$, it follows from the spectral theorem that the v_j form an orthonormal basis of $\mathcal{N}^+\langle\dot{+}\rangle\mathcal{N}^-$, and hence in light of Proposition 6.6, a basis of $\mathcal{N}^+\langle\dot{+}\rangle\mathcal{N}^-$ with respect to the topology induced in this space by the norm $\|\ \|$.

Next let $u \in \mathcal{H}^\# = L^2(\Omega^\#)$ and let $f(x) = u(x)$ for $x \in \Omega^\#$, $f(x) = 0$ for $x \in \Omega\backslash\Omega^\#$. If $0 \in \sigma(A)$ and $M \neq 0$, then it follows from Theorems 3.4-5 and Proposition 6.4 that $f = f_0 + f_1 + f_2$, where $f_0 \in M_0$, $f_1 \in M$, and $f_2 \in \mathcal{H}_1$. If $0 \in \rho(A)$ (resp. if $M = 0$), then we shall still write $f = f_0 + f_1 + f_2$, except for this case we are to take $f_0 = 0$

(resp. $f_1 = 0$). Now given any $\epsilon > 0$, we know from Proposition 6.4 that there is a $g \in V_1$ such that $\|f_2 - g\| < \epsilon$. Hence, since it follows from the foregoing arguments that $g = g_0 + g^+ + g^-$, where $g_0 \in N(K_1) \subset N(T), g^+ \in \mathcal{N}^+$, and $g^- \in \mathcal{N}^-$, we see that if $\| \ \|^\#$ denotes the norm in $\mathcal{H}^\#$, then

$$\left\|u - f_0 - f_1 - g^+ - g^-\right\|^\# = \left\|u - f_0 - f_1 - g\right\|^\# \leq \left\|f - f_0 - f_1 - g\right\| < \epsilon.$$

Thus it follows that the restrictions of the eigenvectors and associated vectors of the problem (2.1) to $\Omega^\#$ form a complete set in $\mathcal{H}^\#$. Moreover, we can now argue as we did in the last paragraph of the proof of Theorem 5.2 to show that the same result also holds in $L^2(\Omega^\#; |\omega(x)|dx)$, and this completes the proof of the theorem.

In the proof of Theorem 6.7 we used the fact that $N(T) \cap V$ was not degenerate with respect to the inner product $B(\ ,\ )$ to show that K_1 had an invariant maximal negative subspace which was not degenerate, and it was precisely this result which enabled us to establish the theorem. We shall now show that this result is false when $N(T) \cap V$ is degenerate, and hence in order to prove the analogue of Theorem 6.7 for this case, different methods will have to be employed. Accordingly, let us suppose that $B(\ ,\ )$ is degenerate on $N(T) \cap V$, or equivalently (see Remark 6.8), that $\mathcal{L} = N(K_1)$ is a degenerate subspace of V_1 (note that for this case to occur we must have $|I_\omega^2| > 0$ and $\kappa_1 > 0$). Then it follows from [**47**, Theorem 3.1, p.103] that $\mathcal{L}$ admits the fundamental decomposition

$$\mathcal{L} = \mathcal{L}_0 \langle\dot{+}\rangle \mathcal{L}_+ \langle\dot{+}\rangle \mathcal{L}_- \tag{6.15}$$

with neutral component $\mathcal{L}_0$, positive definite component $\mathcal{L}_+$, and negative definite component $\mathcal{L}_-$ such that $\dim \mathcal{L}_0 > 0$ and $0 < \dim(\mathcal{L}_0 \langle\dot{+}\rangle \mathcal{L}_-) \leq \kappa_1$ (see [**47**, Lemma 1.1, p.184]).

Proposition 6.7. If $\mathcal{L}$ is a degenerate subspace of V_1, then every maximal negative subspace of V_1 which is invariant under K_1 must be degenerate.

Proof. Suppose that K_1 has an invariant maximal negative subspace $\mathcal{L}^\#$ which is not degenerate. Then we may use the fact that $\sigma(K_1|\mathcal{L}^\#)$ consists precisely of the point $\lambda = 0$ to conclude that $\mathcal{L}^\#$ is negative definite and $\mathcal{L}^\# \subset \mathcal{L}$. Hence it follows from (6.15) and arguments similar to those used in the proof of Proposition 3.3 that $\dim \mathcal{L}^{\#} \leq \dim \mathcal{L}_{-}$. Since $\dim \mathcal{L}^{\#} = \kappa_1$ [**47**, Lemma 1.2, p.184] and $\dim \mathcal{L}_{-} < \kappa_1$, we arrive at a contradiction.

Still fixing our attention upon (6.15), we know from [**47**, Theorem 2.1, p.165 and Theorem 3.2, p.169] that K_1 has an invariant maximal negative subspace $\mathcal{L}^{-}$ containing $\mathcal{L}_0 \langle \dot{+} \rangle \mathcal{L}_{-}$ and an invariant maximal positive subspace $\mathcal{L}^{+}$ which are orthogonal companions of each other. Moreover, $\mathcal{L}^{-}$ is a degenerate subspace of V_1, $\sigma(K_1|\mathcal{L}^{-})$ consists precisely of the point $\lambda = 0$, and $\mathcal{L}^{-}$ is precisely the principal subspace of $K_1|\mathcal{L}^{-}$ corresponding to the eigenvalue $\lambda = 0$. Let $p(\lambda)$ denote the minimum polynomial of $K_1|\mathcal{L}^{-}$. Then $p(\lambda) = \lambda^{\ell}$, where $1 \leq \ell \leq \kappa_1$. Also, since $\langle u, K_1^{\ell} v\rangle = 0$ for $u \in \mathcal{L}^{-}$ and $v \in V_1$, we see that $K_1^{\ell} V_1 \subset \mathcal{L}^{+}$, and hence it follows that $K_1^{2\ell}$ is a positive operator in V_1 (i.e., $\langle K_1^{2\ell} v, v\rangle \geq 0$ for $v \in V_1$). Referring to the statements preceding Theorem 4.2 and to Definition 4.3 for terminology and bearing in mind Remark 4.2, we have next

Theorem 6.8. Suppose that $N(T) \cap V$ is degenerate with respect to the inner product $B(\ ,\ )$. Suppose also that the system (1.1-4) satisfies the condition $(C_{8\ell-1})$, where ℓ is the integer defined above. Then the problem (2.1) has infinitely many positive and infinitely many negative eigenvalues. Moreover, the eigenvectors of (2.1) together with their associated vectors, if any, are complete in $L^2(\Omega^{\#})$ and in $L^2(\Omega^{\#}; |\omega(x)|dx)$.

Proof. Let us fix our attention upon a particular fundamental decomposition of the Pontrjagin space V_1 and let J denote the fundamental symmetry corresponding to this decomposition. Then when V_1, considered only as a vector space, is equipped with the inner product $\langle\ ,\ \rangle_J = \langle J.,.\rangle$, it becomes a Hilbert space and the Hilbert topology of V_1 induced by the norm $\langle u,u\rangle_J^{1/2}$ is precisely the strong topology of V_1. If we now consider V_1 as such a Hilbert space, then it follows that $K_1^{2\ell}$ is a compact operator in V_1 which is strongly symmetrisable with respect to the bounded selfadjoint operator J (see the proof of Theorem 4.1 for terminology) and $S = JK_1^{2\ell} \geq 0$. Futhermore, it follows from (4.7), Theorem 6.1, and Remarks 6.3, 6.7 that the characteristic values and characteristic vectors of K_1 may be arranged into sequences $\{\mu_j\}_1^{\infty}$ and $\{u_j\}_1^{\infty}$, respectively, where each characteristic value is repeated according to its multiplicity, $0 < |\mu_1| \leq |\mu_2| \leq \ldots, (u_j, u_k)_T = \delta_{jk} sgn \mu_j, \langle u_j, u_k\rangle = \delta_{jk}|\mu_j|$, and δ_{jk} is the Kronecker delta.

Let $\phi \in C_0^{\infty}(\Omega^{\#})$, put $f(x) = \phi(x)$ for $x \in \Omega^{\#}, f(x) = 0$ for $x \in \Omega \backslash \Omega^{\#}$, and in Ω let us also define the functions $\{f_j\}_1^{4\ell}$ recursively by putting $f_j(x) = (\omega(x))^{-1}(Lf_{j-1})(x)$ for $x \in \Omega^{\#}$, $f_j(x) = 0$ for $x \in \Omega\backslash\Omega^{\#}$, where $f_0(x) = f(x)$. In light of Definition 2.1 and

Theorem 2.4, it is easy to see that $f_j \in D(A)$ and $Af_j = Tf_{j+1}$ for $j = 0, \ldots, (4\ell - 1)$, while $f_{4\ell} \in V$. Moreover, it follows from Theorems 3.4-6 and Proposition 6.4 that we may write $f_0 = z_0 + w_0$, where $z_0 \in M_0(\dot{+})M$ and $w_0 \in V_1$, while Theorems 3.4-5, 6.4-5, and Remark 6.3 show that if $M_0(\dot{+})M \neq 0$, then there exist the vectors $\{z_j\}_1^{4\ell}$ contained in $M_0(\dot{+})M$ such that $Az_j = Tz_{j+1}$ for $j = 0, \ldots, (4\ell - 1)$. Putting $z_j = 0$ for $j = 1, \ldots, 4\ell$ if $M_0(\dot{+})M = 0$, let $w_j = f_j - z_j$ for $j = 1, \ldots, 4\ell$. Then it follows that $w_j \in V_1$ for $j = 0, \ldots, 4\ell$, while $w_j \in D(A)$ and $Aw_j = Tw_{j+1}$ for $j = 0, \ldots, (4\ell - 1)$. We conclude from this last result and [**154**, Theorem 10, p.422] that

$$w_0 = K_1^{4\ell} w_{4\ell} = \sum_{j=1}^{\infty} \langle Jw_0, u_j \rangle_J u_j. \tag{6.16}$$

Now if $u \in \mathcal{H}^\# = L^2(\Omega^\#)$ and ϵ is any positive number, then we can choose $\phi \in C_0^\infty(\Omega^\#)$ such that $\|u - \phi\|^\# < \epsilon/2$, where $\| \ \|^\#$ denotes the norm in $\mathcal{H}^\#$, and hence it follows from Proposition 6.6 and (6.16) that

$$\left\| u - z_0 - \sum_{j=1}^{p} \langle Jw_0, u_j \rangle_J u_j \right\|^\# < 2^{-1}\epsilon + \left\| f_0 - z_0 - \sum_{j=1}^{p} \langle Jw_0, u_j \rangle_J u_j \right\| < \epsilon$$

if p is sufficiently large. This shows that the restriction of the eigenvectors and associated vectors of the problem (2.1) to $\Omega^\#$ form a complete set in $\mathcal{H}^\#$, while arguments similar to those used in the last paragraph of the proof of Theorem 5.2 show that this is also the case in $L^2(\Omega^\#; |\omega(x)|dx)$.

It remains only to prove that K_1 has infinitely many positive and infinitely many negative characteristic values. To this end let us observe from [**154**, Theorem 10, p.422] and (4.7) that for $u \in V_1^\# = R(K_1^{4\ell})$ we have $u = \sum_{j=1}^{\infty} \mu_j (u, u_j)_T u_j$ and $(u, u)_T = \sum_{j=1}^{\infty} \mu_j^2 |(u, u_j)_T|^2 sgn \mu_j$. Hence if we now consider $V_1^\#$ only as a vector space and equip it with the inner product $(\ , \ )_T$, then it follows that $V_1^\#$ is positive definite if all the μ_j are positive and the orthogonal direct sum of a finite dimensional negative definite subspace and a positive definite subspace if all but a finite number of the μ_j are positive. Thus if we suppose that at most a finite number of the μ_j are negative and let $V^\#$ denote the vector space $M_0(\dot{+})M(\dot{+})V_1^\#$ equipped with the inner product $(\ , \ )_T$, then it follows from our foregoing remarks concerning $V_1^\#$ and the fact that the non-degenerate subspaces M_0 and M are decomposable (see the proof of Theorem 6.7) that $V^\#$ is either positive definite or the orthogonal direct sum of a finite dimensional negative definite subspace and a positive definite subspace. On the other hand, it is a direct consequence of the results of the previous paragraph that $V^{\#}$ contains negative definite subspaces of arbitrary large dimensions, and hence we may argue as we did in the proof of Proposition 3.3 to arrive at a contradiction. Consequently, infinitely many of the μ_j are negative, and similarly we can show that infinitely many are positive to complete the proof of the theorem.

6.5. The system (1.1-4)

In this section we are going to use the results of §6.4 in order to establish some information concerning the spectral properties of the system (1.1-4). Accordingly, recalling the definition of a real eigenvalue of (1.1-4) given in §1.3, we observe from Theorems 2.5 and 6.1 that there may now appear eigenvalues of this system which are not real. Observe also from the Sturm theory that $\lambda^\star = (\lambda_1^\star, \lambda_2^\star)$ is a non-real eigenvalue of (1.1-4) if and only if $\lambda_2^\star$ is not real. Hence the relationship between the real eigenvalues of the system (1.1-4) and those of the problem (2.1) is completely described by Theorems 2.5-6. In order to obtain a similar result for non-real eigenvalues, we shall have to establish the analogue of Theorem 2.6 when the eigenvalue μ there is not real. To this end we firstly require

Lemma 6.2. Suppose that $\lambda_j^\dagger = (\nu_j, \mu), j = 1, \ldots, p$, are any p distinct eigenvalues of the system (1.1-4) having second component μ. Then the $\psi^\star(x, \lambda_j^\dagger), j = 1, \ldots, p$, are linearly independent in $\mathcal{H}$.

Proof. When μ is real, the lemma has been established in the first part of the proof of Theorem 2.6. Hence we suppose from now on that μ is not real and let $L_1(y_1)$ denote the differential expression on the left side of (1.1) when $\lambda_1 = 0$ and $\lambda_2 = \mu$. We are now going to prove the lemma by induction. Accordingly, suppose firstly that $p = 2$ and that the lemma is false. Then there exist non-zero constants $\{c_j\}_1^2$ such that $\sum_{j=1}^2 c_j \psi_j(x) = 0$ for $x \in I^2$, where $\psi_j(x) = \psi^\star(x, \lambda_j^\dagger)$. Applying the differential operator $L_1 + \nu_2 A_1$ to this latter equation, we arrive at the conclusion that $c_1(\nu_2 - \nu_1)A_1(x_1)\psi_1(x) = 0$ for $x = (x_1, x_2) \in I^2$. But this implies that $\psi_1(x) \equiv 0$ in I^2, which is not possible. Thus the lemma is true for $p = 2$.

Suppose next that the lemma is true for all $p \leq k$ for some k satisfying $2 \leq k \in \mathbb{N}$, but that the lemma is false for $p = k + 1$. Then there exist constants $\{c_j\}_1^k$, not all zero, such that $\sum_{j=1}^k c_j \psi_j(x) + \psi_{k+1}(x) = 0$ for $x \in I^2$. Applying the differential operator $L_1 + \nu_{k+1} A_1$ to this latter equation, we arrive at the conclusion that $\sum_{j=1}^k c_j(\nu_{k+1} - \nu_j)A_1(x_1)\psi_j(x) = 0$ for $x = (x_1, x_2) \in I^2$. But this implies that there is a non-trivial linear combination of the ψ_j, $j = 1, \ldots, k$, that vanishes identically in I^2, which contradicts the inductive hypothesis. Hence the lemma is true for $p = k+1$, and the proof is thus complete.

Next let us fix $\mu \in \mathbb{C} \setminus \mathbb{R}$ and let $L_1(y_1)$ and $L_2(y_2)$ denote the differential expressions on the left side of (1.1) and (1.3), respectively, when $\lambda_1 = 0$ and $\lambda_2 = \mu$. Let ν be an eigenvalue of the boundary value problem :

$$(L_1 + \lambda A_1)y = 0, \; 0 \leq x_1 \leq 1, \tag{6.17}$$

together with the boundary condition (1.2), and let $\eta_0(x_1)$ be a corresponding eigenfunction. Then it is well known (cf. [**135**, pp.17-20]) that for some $k \in \mathbb{N}$ there may exist functions $\{\eta_j(x_1)\}_1^k$ which together with their first derivatives are absolutely continuous in $0 \leq x_1 \leq 1$ and satisfy the boundary condition (1.2) such that

$$(L_1 + \nu A_1)\eta_j = -A_1 \eta_{j-1} \;\text{ almost everywhere in }\; 0 < x_1 < 1 \tag{6.18}$$

for $j = 1, \ldots, k$. The functions $\{\eta_j\}_1^k$ are said to be associated with the eigenfunction η_0 and k is called the length of the system of associated functions. Note that

Lemma 6.3. If $\{\eta_j(x_1)\}_1^k$ is a system of functions associated with the eigenfunction η_0, then

$$\int_0^1 A_1(x_1)\eta_j(x_1)\eta_0(x_1)dx_1 = 0 \quad \text{for} \quad j = 0, \ldots, (k-1).$$

Proof. Let us denote the equation (6.17) by (6.17†) when $\lambda = \nu$ and $y = \eta_0$. Then if we multiply (6.18) by η_0, subtract from it (6.17†) multiplied by η_j, integrate over [0,1], and appeal to the boundary condition (1.2), then we obtain $\int_0^1 A_1(x_1)\eta_{j-1}(x_1)\eta_0(x_1)dx_1 = 0$, which is the required result.

Lemma 6.4. Let $\{\eta_j(x_1)\}_1^k$ and $\{\eta_j^\dagger(x_1)\}_1^{k_1}$ be systems of functions associated with η_0 such that $k_1 \geq k$. Then there exist scalers $\{c_j\}_0^k$, where $c_0 = 1$, such that

$$\eta_j(x_1) = \sum_{r=0}^{j} c_{j-r}\eta_r^\dagger(x_1) \;\text{ in }\; 0 \leq x_1 \leq 1 \;\text{ for }\; j = 1, \ldots, k.$$

Proof. It follows from (6.17-18) that there exists the scaler c_1 such that $\eta_1 = \eta_1^\dagger + c_1\eta_0$, which proves the lemma if $k = 1$. Hence suppose that $k > 1$, that $1 \leq r < k$, and that there exist the scalers $\{c_j\}_0^r$, with $c_0 = 1$, such that $\eta_j = \sum_{p=0}^{j} c_{j-p}\eta_p^\dagger$ for $j = 1, \ldots, r$. Then we see from (6.18) that $(L_1 + \nu A_1)\eta_{r+1} = -A_1 \sum_{p=0}^{r} c_{r-p}\eta_p^\dagger = (L_1 + \nu A_1)(\sum_{p=0}^{r} c_{r-p}\eta_{p+1}^\dagger)$, and hence it follows that there exists the scaler c_{r+1} such that $\eta_{r+1} = \sum_{p=0}^{r+1} c_{r+1-p}\eta_p^\dagger$. The proof of the lemma is now completed by arguing inductively.

Turning again to the eigenfunction η_0 introduced above, we say that η_0 has multiplicity $p_1^{\#}(\eta_0)$ if there is a system of functions associated with η_0 of length $p_1^{\#}(\eta_0)-1$, but no system of length $p_1^{\#}(\eta_0)$. Note that this definition does not exclude the case $p_1^{\#}(\eta_0) = \infty$; however we shall see in the sequel that we always have $p_1^{\#}(\eta_0) < \infty$. Moreover,

Lemma 6.5. Let $\eta_0^\dagger(x_1)$ be an eigenfunction of the problem (6.17), (1.2) corresponding to the eigenvalue ν which is distinct from $\eta_0(x_1)$. Then $p_1^{\#}(\eta_0^\dagger) = p_1^{\#}(\eta_0)$.

Proof. Let $\{\eta_j\}_1^{p_1^{\#}(\eta_0)-1}$ be a system of functions associated with η_0 of length $p_1^{\#}(\eta_0) - 1$. Then since $\eta_0^\dagger = c\eta_0$ for some constant $c \neq 0$, we see from (6.18) that $\{c\eta_j\}_1^{p_1^{\#}(\eta_0)-1}$ is a system of functions associated with $\eta_0^\dagger$, and hence we conclude that $p_1^{\#}(\eta_0^\dagger) \geq p_1^{\#}(\eta_0)$. By interchanging the roles of η_0 and $\eta_0^\dagger$ in the above arguments, it also follows that $p_1^{\#}(\eta_0) \geq p_1^{\#}(\eta_0^\dagger)$, and this completes the proof of the lemma.

Similarly if ν is an eigenvalue of the boundary value problem

$$(L_2 - \lambda A_2)y = 0,\ 0 \leq x_2 \leq 1, \tag{6.19}$$

together with the boundary condition (1.4), and $\varsigma_0(x_2)$ is a corresponding eigenfunction, then a system of functions associated with ς_0 is defined in an analogous manner to that for η_0 above. In particular, if $\{\varsigma_j\}_1^k$, $k \in \mathbb{N}$, is such a system, then we have

$$(L_2 - \nu A_2)\varsigma_j = A_2\varsigma_{j-1} \text{ almost everywhere in } 0 < x_2 < 1 \tag{6.20}$$

for $j = 1, \ldots, k$ and $\int_0^1 A_2(x_2)\varsigma_j(x_2)\varsigma_0(x_2)dx_2 = 0$ for $j = 0, \ldots, (k-1)$. We define the multiplicity of the eigenfunction ς_0 in an analogous manner to that for η_0 above and denote it by $p_2^{\#}(\varsigma_0)$; and again we have $p_2^{\#}(\varsigma_0) < \infty$ and $p_2^{\#}(\varsigma_0) = p_2^{\#}(\varsigma_0^\dagger)$ if

$\varsigma_0^\dagger$ is an eigenfunction of the problem (6.19), (1.4) corresponding to the eigenvalue ν which is distinct from ς_0. Finally, if $\lambda^\dagger = (\nu, \mu)$ is an eigenvalue of the system (1.1-4) (i.e., if ν is a common eigenvalue of the problems (6.17),(1.2) and (6.19),(1.4)), and with $\eta_0, p_1^\#(\eta_0), \varsigma_0, p_2^\#(\varsigma_0)$ defined as above, if $p_1^\#(\eta_0)$ and $p_2^\#(\varsigma_0)$ are both greater than 1, if $\{\eta_j\}_1^{p_1^\#(\eta_0)-1}$ and $\{\varsigma_j\}_1^{p_2^\#(\varsigma_0)-1}$ are systems of functions associated with η_0 and ς_0, respectively, and if $p = \min\{p_1^\#(\eta_0) - 1, p_2^\#(\varsigma_0) - 1\}$, then the functions $\sum_{j=0}^{\ell} \eta_{\ell-j}(x_1)\varsigma_j(x_2), \ell = 1, \ldots, p$, will be called generalized eigenfunctions of the system (1.1-4) corresponding to the eigenvalue $\lambda^\dagger$. It is important to observe from the definitions of the terms involved that these generalized eigenfunctions, as well as the eigenfunction $\eta_0\varsigma_0$, all belong to the class U of functions defined in the statements immediately preceding Theorem 3.5. It is also important to note that

Lemma 6.6. Suppose that $\lambda^\dagger = (\nu, \mu)$ is an eigenvalue of the system (1.1-4) and that, with the above notation, $p_1^\#(\eta_0)$ and $p_2^\#(\varsigma_0)$ are both greater than 1. Suppose also that $\eta_0^\dagger$ and $\varsigma_0^\dagger$ are eigenfunctions of the problems (6.17),(1.2) and (6.19),(1.4), respectively, corresponding to the eigenvalue ν and that $\{\eta_j^\dagger\}_1^{p_1^\#(\eta_0^\dagger)-1}$ and $\{\varsigma_j^\dagger\}_1^{p_2^\#(\varsigma_0^\dagger)-1}$ are systems of functions associated with $\eta_0^\dagger$ and $\varsigma_0^\dagger$, respectively. Then $p_1^\#(\eta_0^\dagger) = p_1^\#(\eta_0)$, $p_2^\#(\varsigma_0^\dagger) = p_2^\#(\varsigma_0)$, and the linear manifold $\mathcal{H}$ spanned by the vectors $\{\sum_{j=0}^{\ell} \eta_{\ell-j}^\dagger \varsigma_j^\dagger\}_{\ell=0}^{p}$ is precisely $\text{span}\{\sum_{j=0}^{\ell} \eta_{\ell-j}\varsigma_j\}_{\ell=0}^{p}$, where the integer p is defined above.

Proof. The assertions of the lemma concerning multiplicities are an immediate consequence of Lemmas 6.4-5 and their analogues for the problem (6.19),(1.4). Turning to the final assertion and writing p_1 for $p_1^\#(\eta_0)$, p_2 for $p_2^\#(\varsigma_0)$, we may argue as we did in the proof of Lemma 6.4 to show that there exist the scalers $\{c_j\}_0^{p_1-1}, \{d_j\}_0^{p_2-1}$, where $c_0 \neq 0$ and $d_0 \neq 0$, such that

$$\eta_j^\dagger = \sum_{k=0}^{j} c_{j-k}\eta_k \quad \text{for} \quad j = 0, \ldots, (p_1 - 1),$$

$$\varsigma_j^\dagger = \sum_{k=0}^{j} d_{j-k}\varsigma_k \quad \text{for} \quad j = 0, \ldots, (p_2 - 1).$$

Thus if $0 \leq \ell \leq p$, then

$$\sum_{j=0}^{\ell} \eta_{\ell-j}^\dagger \varsigma_j^\dagger = \sum_{j=0}^{\ell} \left(\sum_{r=0}^{\ell-j} c_{\ell-j-r} d_r \right) \sum_{k=0}^{j} \eta_{j-k}\varsigma_k,$$

and hence it follows that $\text{span}\{\sum_{j=0}^{\ell} \eta_{\ell-j}^{\dagger} \varsigma_j^{\dagger}\}_{\ell=0}^{p}$ is a subspace of $\text{span}\{\sum_{j=0}^{\ell} \eta_{\ell-j} \varsigma_j\}_{\ell=0}^{p}$. By interchanging the roles of the $\eta_j^{\dagger}, \varsigma_j^{\dagger}$ and the η_j, ς_j in the above arguments, we obtain the reverse inclusion, and this completes the proof of the lemma.

Remark 6.9. In view of Lemma 6.6, we shall henceforth refer to either of the system of functions $\{\sum_{j=0}^{\ell} \eta_{\ell-j} \varsigma_j\}_{\ell=1}^{p}$ or $\{\sum_{j=0}^{\ell} \eta_{\ell-j}^{\dagger} \varsigma_j^{\dagger}\}_{\ell=1}^{p}$ as the generalized eigenfunctions of (1.1-4) corresponding to the eigenvalue $\lambda^{\dagger}$.

We now come to the first of our main results, namely,

Theorem 6.9. Suppose that μ is a non-real eigenvalue of the problem (2.1) and that $\dim N_{\mu} = \eta_{\mu} \in \mathbb{N}$. Then there are precisely m_{μ} distinct eigenvalues of the system (1.1-4) whose second components are μ, where $1 \leq m_{\mu} \leq n_{\mu}$. Moreover, if we denote these eigenvalues by $\{\lambda_j^{\dagger}\}_1^{m_{\mu}}$, then $\{\psi^{\star}(x, \lambda_j^{\dagger})\}_1^{m_{\mu}}$ is a basis of N_{μ} if $m_{\mu} = n_{\mu}$. On the other hand, if $m_{\mu} < n_{\mu}$, then for at least one j there exist generalized eigenfunctions of the system (1.1-4) corresponding to the eigenvalue $\lambda_j^{\dagger}$, and there is a basis of N_{μ} consisting of $\{\psi^{\star}(x, \lambda_j^{\dagger})\}_1^{m_{\mu}}$ together with the generalized eigenfunctions of the system (1.1-4) corresponding to those eigenvalues $\lambda_j^{\dagger}$ for which such generalized eigenfunctions exist.

Proof. Let $\mathcal{H}_1$ denote the Hilbert space $L^2(0 < x_1 < 1; A_1(x_1)dx_1)$ and let $\mathcal{A}_0$ denote the selfadjoint operator in $\mathcal{H}_1$ generated by the differential expression $-(A_1(x_1))^{-1}(p_1(x_1)y')'$ and the boundary condition (1.2), where $' = d/dx_1$. Then it is well known (cf. [**55**, Chapter 8], [**136**, Chapter V]) that $\mathcal{A}_0$ has compact resolvent and simple eigenvalues $\lambda_1 < \lambda_2 < \ldots$, while arguments somewhat similar to those used in the proof of Theorem 8.4.1 of [**107**, p.404] show that $\lambda_j - \lambda_{j-1} \to \infty$ as $j \to \infty$. Thus if $\mathcal{A}$ denotes the operator in $\mathcal{H}_1$ generated by the differential expression $-(A_1(x_1))^{-1} L_1 y$ and the boundary condition (1.2), where L_1 is defined above (see (6.17)), then we know from [**120**, Theorem 4.15a, p.293] that $\mathcal{A}$ is closed and has compact resolvent, and hence a discrete spectrum. Moreover, the eigenvalues of $\mathcal{A}$ are denumerably infinite in number and, with the possible exception of at most a finite number of them, are all simple.

Let $\theta(x_1, \lambda)$ and $\theta_1(x_1, \lambda)$ denote the solutions of (6.17) satisfying $\theta(0, \lambda) = \sin \alpha_1, p_1(0)\theta'(0, \lambda) = \cos \alpha_1$ and $\theta_1(1, \lambda) = \sin \beta_1, p_1(1)\theta_1'(1, \lambda) = \cos \beta_1$, respectively, and let $\delta(\lambda) = p_1(x_1)W(\theta, \theta_1)(x_1, \lambda)$, where W denotes the Wronskian and $' = d/dx_1$. Then it is clear that the zeros of the entire function $\delta(\lambda)$ are precisely the eigenvalues of the operator $\mathcal{A}$ introduced above, and hence it follows that $\delta(\lambda)$ is not identically zero. Furthermore, it is not difficult to verify that for each λ in $\mathbb{C}$,

$$\delta^{(r)}(\lambda) = \begin{vmatrix} \theta^{(r)}(1,\lambda) & \sin\beta_1 \\ p_1(1)(\theta')^{(r)}(1,\lambda) & \cos\beta_1 \end{vmatrix},$$

$$r\int_0^1 A_1(x_1)(\mathcal{D}^{r-1}\theta)(x_1,\lambda)\theta(x_1,\lambda)dx_1 = \tag{6.21}$$

$$= \begin{vmatrix} \theta^{(r)}(1,\lambda) & \theta(1,\lambda) \\ p_1(1)(\theta')^{(r)}(1,\lambda) & p_1(1)\theta'(1,\lambda) \end{vmatrix},$$

and

$$\Big[(L_1 + \lambda A_1)((1/r!)\mathcal{D}^r\theta)\Big](x_1,\lambda) = -A_1(x_1)\Big[(1/(r-1)!)\mathcal{D}^{r-1}\theta\Big](x_1,\lambda)$$

$$\text{almost everywhere in } 0 < x_1 < 1$$

for $r \geq 0$, where ${}^{(r)} = d^r/d\lambda^r$, $\mathcal{D} = \partial/\partial\lambda$, and $\mathcal{D}^{-1}\theta = 0$. Thus in particular, the zero λ_0 of $\delta(\lambda)$ is simple if and only if $\int_0^1 A_1(x_1)\big(\theta(x_1,\lambda_0)\big)^2 dx_1 \neq 0$. Now let $\{\Lambda_j\}_0^\infty$ denote the zeros of $\delta(\lambda)$ arranged in increasing order of magnitude with each zero being repeated according to its multiplicity. If Λ_r is a simple zero of $\delta(\lambda)$, then let $\eta_r(x_1) = c_r^{1/2}\theta(x_1,\Lambda_r)$, where $c_r^{-1} = \int_0^1 A_1(x_1)\ \big(\theta(x_1,\Lambda_r)\big)^2 dx_1$ and the principal determination of the square root is taken. It is clear from (6.21) and Lemma 6.3 that the eigenfunction η_r corresponding to the eigenvalue Λ_r of the problem (6.17),(1.2) has multiplicity 1, and hence we conclude from Lemma 6.5 that Λ_r is a simple eigenvalue of $\mathcal{A}$ and η_r a corresponding eigenvector. If Λ_r is a zero of $\delta(\lambda)$ of order $p > 1$ and $\Lambda_r = \Lambda_{r+1} = \ldots = \Lambda_{r+p-1}$, then let $\eta_r(x_1) = \theta(x_1,\Lambda_r)$. It follows immediately from (6.21) and Lemmas 6.3-4 that for this case the eigenfunction η_r of (6.17),(1.2) for the eigenvalue Λ_r has multiplicity p and for $j = 1,\ldots,(p-1)$, the $\eta_{r+j}(x_1) = (1/j!)(\mathcal{D}^j\theta)(x_1,\Lambda_r)$ is a system of functions of length $(p-1)$ associated with η_r. Since it follows from (6.18), with ν replaced by Λ_r and η_j by η_{r+j}, that $\{\eta_{r+j}\}_{j=0}^{p-1}$ is a linearly independent set in $\mathcal{H}_1$, it is now a simple matter to deduce from (6.18), Lemma 6.5, and the assertions made in the proof of Lemma 6.6 that p is precisely the algebraic multiplicity of the eigenvalue Λ_r of $\mathcal{A}$ and that span$\{\eta_{r+j}\}_{j=0}^{p-1}$ is the corresponding principal subspace. Thus we have shown that the zeros of $\delta(\lambda)$ are all simple with the possible exception of at most a finite number of them, while we know from [**120**, Theorem 4.15a, p.293] that the sequence $\{\eta_j\}_0^\infty$ is an unconditional basis of $\mathcal{H}_1$.

We are now going to construct the sequence $\{\psi_j\}_0^\infty$ which is biorthogonal to $\{\eta_j\}_0^\infty$ in $\mathcal{H}_1$ (we refer the reader to [**98**, Chapter VI] for information regarding biorthogonal sequences). Accordingly, let $L_1^\star$ denote the differential expression which is obtained from L_1 by replacing μ by $\overline{\mu}$. Then it is well known (cf. [**120**, Theorem 2.5 and examples 1.7, 1.36, and 2.16, Chapter VI]) that the closed operator in $\mathcal{H}_1$ generated by the differential expression $-(A_1(x_1))^{-1}L_1^\star$ and the boundary condition (1.2) is precisely $\mathcal{A}^\star$, the adjoint of $\mathcal{A}$ in $\mathcal{H}_1$, and that $D(\mathcal{A}) = D(\mathcal{A}^\star)$. Moreover, $\mathcal{A}^\star$ has a discrete spectrum, with the eigenvalues of $\mathcal{A}^\star$ being precisely the complex conjugates of those of $\mathcal{A}$, while if λ is an eigenvalue of $\mathcal{A}$ and $\mathcal{G}_\lambda$ denotes the principal subspace of $\mathcal{A}$ for the eigenvalue λ, then $\dim \mathcal{G}_\lambda = \dim \mathcal{G}^\star_{\overline{\lambda}}$, where $\mathcal{G}^\star_{\overline{\lambda}}$ denotes the principal subspace of $\mathcal{A}^\star$ for the eigenvalue $\overline{\lambda}$ (see [**120**, p.184]). Arguments similar to those used in the proofs of Theorem 6.3 and Lemma 6.1 also show that if λ is an eigenvalue of $\mathcal{A}$, then $\mathcal{G}^\star_{\overline{\lambda}}$ is composed of those vectors in $\mathcal{H}_1$ which are precisely the complex conjugates of the vectors of $\mathcal{G}_\lambda$ and that $\overline{(\mathcal{A}-\lambda I)u} = (\mathcal{A}^\star - \overline{\lambda} I)\overline{u}$ for $u \in \mathcal{G}_\lambda$, where $\overline{u}(x_1)$ denotes the complex conjugate of $u(x_1)$, while it follows from [**120**, p.184 and Theorem 4.15a, p.293] that if λ_1 and λ_2 are any two distinct eigenvalues of $\mathcal{A}$, then $\mathcal{G}_{\lambda_1}$ and $\mathcal{G}^\star_{\overline{\lambda}_2}$ are orthogonal in $\mathcal{H}_1$. Hence if Λ_r is a simple zero of $\delta(\lambda)$ and if we put $\psi_r = \overline{\eta}_r$, then $(\eta_j, \psi_r)_1 = \delta_{jr}$, where $(\ ,\ )_1$ denotes the inner product in $\mathcal{H}_1$ and δ_{jr} denotes the Kronecker delta. If Λ_r is a zero of $\delta(\lambda)$ of order $p > 1$ and $\Lambda_r = \Lambda_{r+1} = \ldots = \Lambda_{r+p-1}$, then we have $(\eta_j, \overline{\eta}_{r+k})_1 = 0$ for $k = 0, \ldots, (p-1)$ provided that $j \neq r+k$, $k = 0, \ldots, (p-1)$. Furthermore, if we observe from (6.18), with ν and η_j replaced by Λ_r and η_{r+j}, respectively, that $(\eta_{r+j}, \overline{\eta}_{r+k})_1 = (\eta_{r+j+1}, \overline{\eta}_{r+k-1})_1$ for $0 \le j < p-1$, $1 \le k \le p-1$, and if we put $I_{jk} = (\eta_{r+j}, \overline{\eta}_{r+k})_1$ for $0 \le j, k \le p-1$, then it follows immediately from (6.21) that I_{jk} is constant for $j+k =$ constant and $I_{jk} = 0$ if $j+k < p-1$, $I_{jk} \neq 0$ if $j+k = p-1$. This shows that there exist the unique constants $\{c_{rj}\}_{j=0}^{p-1}$, with $c_{r0} \neq 0$, such that if we put $w_{r+j} = \sum_{k=0}^{j} c_{r,j-k}\eta_{r+k}$ for $j = 0, \ldots, (p-1)$, then $(\eta_{r+j}, \overline{w}_{r+k})_1$ is 0 or 1 according to whether $j+k \neq p-1$ or $j+k = p-1$. We note from Lemmas 6.3-5 and (6.21) that the eigenfunction w_r of the problem (6.17),(1.2) corresponding to the eigenvalue Λ_r has multiplicity p and that $\{w_{r+j}\}_{j=1}^{p-1}$ is a system of functions associated with w_r. Finally, if we put $\psi_{r+k} = \overline{w}_{r+p-1-k}$ for $k = 0, \ldots, (p-1)$, then we have $(\eta_j, \psi_{r+k})_1 = \delta_{j,r+k}$. Thus we have shown that the sequence $\{\psi_j\}_0^\infty$, as constructed above, is biorthogonal to $\{\eta_j\}_0^\infty$ in $\mathcal{H}_1$.

Let $\{u_s\}_1^{n_\mu}$ be a basis of N_μ and let us fix our attention upon a particular u_s and write u for u_s. Then we may argue as we did in the proof of Theorem 2.6 to show that by modifying u on a set of measure zero and extending u to all of I^2 by continuity, we may henceforth suppose that $u \in C^1(I^2)$ and satisfies the boundary conditions (1.2) and (1.4) (see §2.2 for terminology). Hence, in light of the foregoing results, we now have

$$u(.,x_2) = \sum_{r=0}^{\infty}(u(.,x_2),\psi_r)_1\eta_r(.) \text{ in } \mathcal{H}_1 \text{ for } 0 \le x_1 \le 1. \tag{6.22}$$

Now suppose that Λ_r is a simple zero of $\delta(\lambda)$ and let

$$\chi_r(x_2) = \int_0^1 A_1(x_1)\eta_r(x_1)u(x_1,x_2)dx_1.$$

Then it is an immediate consequence of the properties of u cited above that χ_r is of class C^1 in $0 \le x_2 \le 1$ and satisfies the boundary condition (1.4), while arguments similar to those used in the proof of Theorem 2.6 show that $d\chi_r/dx_2$ is absolutely continuous in $0 \le x_2 \le 1$ and $(L_2 - \Lambda_r A_2)\chi_r = 0$ almost everywhere in $0 < x_2 < 1$, where L_2 is defined above (see (6.19)). Thus we conclude that if χ_r is not identically zero, then $\eta_r\chi_r$ is an eigenfunction of the system (1.1-4) corresponding to the eigenvalue (Λ_r,μ). Suppose next that Λ_r is a zero of $\delta(\lambda)$ of order $p > 1, \Lambda_r = \Lambda_{r+1} = \ldots = \Lambda_{r+p-1}$, and

$$\chi_{r+j}(x_2) = \int_0^1 A_1(x_1)w_{r+j}(x_1)u(x_1,x_2)dx_1$$

for $j = 0,\ldots,(p-1)$. Then recalling the properties of the w_{r+j} cited above, we can again show that for each j, χ_{r+j} is of class C^1 and $d\chi_{r+j}/dx_2$ is absolutely continuous in $0 \le x_2 \le 1, \chi_{r+j}$ satisfies the boundary condition (1.4), and

$$(L_2 - \Lambda_r A_2)\chi_{r+j} = A_2\chi_{r+j-1} \text{ almost everywhere in } 0 < x_2 < 1, \tag{6.23}$$

where $\chi_{r-1} = 0$. Thus if the χ_{r+j} are not all identically equal to zero, and $i, 0 \le i \le p-1$, denotes the smallest value of j for which χ_{r+j} is not identically equal to zero, then $\eta_r\chi_{r+i}$ is an eigenfunction of the system (1.1-4) corresponding to the eigenvalue (Λ_r,μ).

We know from Lemma 6.2 that there are at most n_μ distinct eigenvalues of the system (1.1-4) having second component μ, while it is clear from the foregoing arguments and (6.22) that there is at least one. If we denote these eigenvalues by $\lambda_j^\dagger = (\nu_j,\mu)$, $j = 1,\ldots,m_\mu$, then obviously χ_r must be identically zero if Λ_r is a simple zero of $\delta(\lambda)$ not coinciding with any of the ν_j. If Λ_r is a zero of $\delta(\lambda)$ of order $p > 1$, if $\Lambda_r = \Lambda_{r+1} = \ldots = \Lambda_{r+p-1}$, and if Λ_r does not coincide with any of the ν_j, then it follows from (6.23) that χ_{r+j} is identically zero for $j = 0,\ldots,(p-1)$. Hence we conclude from (6.22) that

$$u(x) = \sum_r Z_r(x) \text{ for } x \in \Omega, \tag{6.24}$$

where the summation is over those m_μ distinct values of r for which : (1) Λ_r is a simple zero of $\delta(\lambda)$ which coincides with a ν_j, if such simple zeros exist, (2) Λ_r is a zero of $\delta(\lambda)$ of order $p > 1$, $\Lambda_r = \Lambda_{r+1} = \ldots = \Lambda_{r+p-1}$, and Λ_r coincides with a ν_j, if such multiple zeros exist, and where for an r for which case (1) holds we have $Z_r = \eta_r \chi_r$, while for an r for which case (2) holds we have $Z_r = \sum_{k=0}^{p-1} \eta_{r+p-1-k} \chi_{r+k}$.

Let $\varsigma(x_2, \lambda)$ denote the solution of (6.19) satisfying $\varsigma(0, \lambda) = \sin \alpha_2, p_2(0)\varsigma'(0, \lambda) = \cos \alpha_2$, where $' = d/dx_2$. If Λ_r is a simple zero of $\delta(\lambda)$ and $\Lambda_r = \nu_j$ for some j, then χ_r may or may not be identically zero, and in the former case Z_r does not appear in the summation on the right side of (6.24), while in the latter case Z_r is a scalar multiple of the eigenfunction $\eta_r(x_1)\varsigma(x_2, \nu_j)$ of the system (1.1-4). Let us next fix our attention upon the case where Λ_r is a zero of $\delta(\lambda)$ of order $p > 1, \Lambda_r = \Lambda_{r+1} = \ldots = \Lambda_{r+p-1}$, and $\Lambda_r = \nu_j$ for some j. If χ_{r+k} is identically zero for $k = 0, \ldots, (p-1)$, then Z_r does not appear in the summation on the right side of (6.24). If not all of the χ_{r+k} are identically zero, then let i denote the smallest value of k for which $\chi_{r+k} \not\equiv 0$. If $i = p - 1$, then we see from (6.23) that Z_r is a scalar multiple of the eigenfunction $\eta_r \varsigma_0$ of (1.1-4), where $\varsigma_0(x_2) = \varsigma(x_2, \nu_j)$. If $0 \le i < p - 1$, then it follows from (6.20) and (6.23) that χ_{r+i} is an eigenfunction of the problem (6.19),(1.4) corresponding to the eigenvalue ν_j and that $\{\chi_{r+k}\}_{k=i+1}^{p-1}$ is a system of functions associated with χ_{r+i}. We conclude immediately from the analogues of (6.21) and Lemmas 6.3-4 for the problem (6.19),(1.4) that ς_0 is an eigenfunction of (6.19),(1.4) corresponding to the eigenvalue ν_j whose multiplicity is at least equal to $p - i$ and that $\varsigma_k(x_2) = (1/k!)(\mathcal{D}^k \varsigma)(x_2, \nu_j)$, $k = 1, \ldots, (p - i - 1)$, is a system of functions associated with ς_0, while arguments similar to those used in the proof of Lemma 6.4 show that

$$\chi_{r+k} = \sum_{\ell=0}^{k-i} d_{r,k-i-\ell} \varsigma_\ell \quad \text{for} \quad k = i, \ldots, (p-1),$$

where the d_{rk} are constants and $d_{r0} \neq 0$. Thus it follows that $Z_r = \sum_{\ell=0}^{p-i-1} d_{r\ell} Z_{r\ell}$, where $Z_{r\ell} = \sum_{k=0}^{p(\ell)} \eta_{r+p(\ell)-k} \varsigma_k$ and $p(\ell) = p - i - 1 - \ell$, and hence Z_r is a linear combination of the eigenfunction $\eta_r \varsigma_0$ and the generalized eigenfunctions $\{Z_{r\ell}\}_{\ell=0}^{p-i-2}$ of (1.1-4) corresponding to the eigenvalue (ν_j, μ).

Let us now investigate this last case in greater detail. Suppose that the eigenfunction ς_0 of the problem (6.19),(1.4) corresponding to the eigenvalue ν_j has multiplicity $p^* \ge p - i$. Then it follows from the analogues of (6.21) and Lemmas 6.3-4 for the problem (6.19),(1.4) that $\{\varsigma_k\}_1^{p^*-1}$ is a system of functions associated with ς_0, where the ς_k are defined in the previous paragraph. Let $p_j = \min\{p - 1, p^* - 1\}$ and

$\Phi_{\ell k}(x) = \eta_{r+\ell}(x_1)\varsigma_k(x_2)$ for $0 \leq \ell, k \leq p_j$. Then it is clear from the definitions of the terms involved that $\Phi_{\ell k}$ belongs to the class U of functions on I^2 defined in the statements preceding Theorem 3.5, and hence it follows from Theorem 2.4 that $\Phi_{\ell k} \in D(A)$. Moreover, we have

$$(L\Phi_{\ell k})(x) = \mu\omega(x)\Phi_{\ell k}(x) - A_2(x_2)((L_1 + \nu_j A_1)\Phi_{\ell k})(x) -$$

$$- A_1(x_1)((L_2 - \nu_j A_2)\Phi_{\ell k})(x) \tag{6.25}$$

$$= \mu\omega(x)\Phi_{\ell k}(x) + A_1(x_1)A_2(x_2)\Big[\Phi_{\ell-1,k}(x) - \Phi_{\ell,k-1}(x)\Big]$$

almost everywhere in Ω, where $\Phi_{\ell k}(x) = 0$ if $\ell = -1$ or $k = -1$. Hence if we put $R_{j\ell}(x) = \sum_{k=0}^{\ell} \Phi_{\ell-k,k}(x)$ for $\ell = 0, \ldots, p_j$, then for each ℓ, $R_{j\ell} \in D(A)$ and, in light of (6.25), we also see that $(LR_{j\ell})(x) = \mu\omega(x)R_{j\ell}(x)$ almost everywhere in Ω. Thus it follows from Theorem 2.4 that $R_{j\ell} \in N_\mu$ for $\ell = 0, \ldots, p_j$.

We now construct a basis of N_μ in the following way. If $\nu_j = \Lambda_r$ and Λ_r is a simple zero of $\delta(\lambda)$, then put $p_j = 0$ and $R_{j0}(x) = \theta(x_1, \nu_j)\varsigma(x_2, \nu_j)$ (see the definitions following (6.21)). If $\nu_j = \Lambda_r$, if Λ_r is a zero of $\delta(\lambda)$ of order $p > 1$, if $\Lambda_r = \Lambda_{r+1} = \ldots = \Lambda_{r+p-1}$, and referring to the previous paragraph for notation, if ς_0 is an eigenfunction of the problem (6.19),(1.4) corresponding to the eigenvalue ν_j of multiplicity $p^\star$, then let $p_j = \min\{p-1, p^\star - 1\}$, $R_{j0}(x) = \eta_r(x_1)\varsigma_0(x_2)$ if $p_j = 0$, and let the $R_{j\ell}(x), \ell = 0, \ldots, p_j$, be defined as in the previous paragraph if $p_j > 0$. Then the $R_{j\ell}, j = 1, \ldots, m_\mu, \ell = 0, \ldots, p_j$, are contained in N_μ, and moreover, we may use the fact that the sequences $\{\eta_r\}_0^\infty$ and $\{\psi_r\}_0^\infty$ are biorthogonal in $\mathcal{H}_1$ to show that the $R_{j\ell}$ are linearly independent in $\mathcal{H}$. Finally, since we have shown that each member of the basis $\{u_s\}_1^{n_\mu}$ of N_μ is a linear combination of the $R_{j\ell}$, it follows that $\{R_{j\ell}\}, j = 1, \ldots, m_\mu$, $\ell = 0, \ldots, p_j$, is precisely the basis of N_μ whose existence is asserted in the theorem.

Remark 6.10. Referring again to Theorem 6.9, suppose that $1 \leq j \leq m_\mu$. Then it follows from the proof of the theorem as well as from the analogues of (6.21) and Lemma 6.3 for the problem (6.19),(1.4) that in order for there to exist generalized eigenfunctions of the system (1.1-4) corresponding to the eigenvalue $\lambda_j^\dagger$, it is necessary and sufficient that $\int_0^1 A_r(x_r)\phi_r^2(x_r, \lambda_j^\dagger)dx_r = 0$ for $r = 1, 2$ (see §1.3 for notation).

Remark 6.11. Suppose that in Theorem 6.9, $m_\mu > 1$, $j \neq k$, and u (resp. v) denotes either the eigenfunction $\psi^\star(x, \lambda_j^\dagger)$ (resp. $\psi^\star(x, \lambda_k^\dagger)$) or a generalized eigenfunction of

(1.1-4) corresponding to the eigenvalue $\lambda_j^\dagger$ (resp. $\lambda_k^\dagger$) if any such functions exist. Then it follows from the properties of the η_r cited in the proof of Theorem 6.9 as well as from the analogous results concerning the eigenfunctions and associated functions of the problem (6.19),(1.4) that $(u, \overline{v})_T = 0$.

We are now in a position to present some further results concerning the system (1.1-4). Accordingly, if $\lambda = (\lambda_1, \lambda_2)$ and $\nu = (\nu_1, \nu_2)$ are any two eigenvalues of (1.1-4), then we say that λ is the complex conjugate of ν if $\lambda_i = \overline{\nu}_i$ for $i = 1, 2$.

Theorem 6.10. The eigenvalues of the system (1.1-4) form a denumerably infinite subset of $\mathbb{C}^2$ having no finite points of accumulation. Moreover, these eigenvalues are all real, with the possible exception of at most a finite number of them, and if non-real eigenvalues do appear, then they must occur in complex conjugate pairs and the total number of distinct non-real eigenvalues does not exceed 2κ.

Proof. All the assertions of the theorem except the one concerning complex conjugate pairs follow from Theorems 2.5-6, 6.1-2, and 6.9. On the other hand, we have seen in the proof of Theorem 6.9 that if ν is an eigenvalue of the problem (6.17),(1.2) then $\overline{\nu}$ is an eigenvalue of the problem (6.17)*,(1.2), where (6.17)* denotes the equation obtained from (6.17) when L_1 is replaced by $L_1^\star$. Since a similar result holds for the problem (6.19),(1.4), we conclude that the non-real eigenvalues of the system (1.1-4) must occur in complex conjugate pairs.

Theorem 6.11. Let E denote the set consisting of those distinct real eigenvalues $\lambda^\dagger = (\lambda_1^\dagger, \lambda_2^\dagger)$ of the system (1.1-4) for which $\lambda_2^\dagger \neq 0$ and $(v, v)_T = 0$, where $v = \psi^\star(x, \lambda^\dagger)$. Then the cardinality of E does not exceed κ.

Proof. We know from Theorem 2.5 and the proof of Theorem 6.1 that if $\lambda^\dagger = (\lambda_1^\dagger, \lambda_2^\dagger) \in E$, then $\lambda_2^\dagger$ is a characteristic value of K_0 and $\psi^\star(x, \lambda^\dagger)$ a corresponding characteristic vector. Hence it follows from Theorem 1.2 and (4.7) that the $\psi^\star(x, \lambda^\dagger)$, $\lambda^\dagger \in E$, are linearly independent and span a neutral subspace of the Pontrjagin space V_0. The proof of the theorem is now completed by appealing to [**47**, Lemma 1.1., p.184].

Theorem 6.12. Let E denote the set consisting of those distinct real eigenvalues $\lambda^\dagger = (\lambda_1^\dagger, \lambda_2^\dagger)$ of the system (1.1-4) for which $\lambda_2^\dagger \neq 0$ and $sgn(v, v)_T \neq sgn\lambda_2^\dagger$, where $v(x) = \psi^\star(x, \lambda^\dagger)$. Then the cardinality of E does not exceed κ.

Proof. We may argue as in the proof of Theorem 6.11 to show that the $\psi^*(x, \lambda^\dagger)$, $\lambda^\dagger \in E$, are linearly independent and span a negative subspace of the Pontrjagin space V_0. Hence the theorem follows from [**47**, Lemma 1.1, p.184].

Referring next to Theorems 6.7-8 for terminology we also have

Theorem 6.13. Suppose that the system (1.1-4) satisfies the condition $(C_{8\ell-1})$ if $N(T) \cap V$ is degenerate with respect to the inner product $B(\ ,\ )$. Then the system (1.1-4) has infinitely many eigenvalues whose second components are positive and infinitely many whose second components are negative.

Proof. The assertions of the theorem follow immediately from Theorems 2.6 and 6.7-8.

Theorem 6.14. Suppose that μ is a non-real eigenvalue of the problem (2.1) and let $\{\lambda_j^\dagger\}_1^{m_\mu}$ denote the distinct eigenvalues of the system (1.1-4) whose second components are μ. Let $u_j(x) = \psi^*(x, \lambda_j^\dagger)$ for $j = 1, \ldots, m_\mu$. Then μ is a semi-simple eigenvalue of the problem (2.1) and $\{u_j\}_1^{m_\mu}$ is a basis of M_μ if and only if $(u_j, \overline{u}_j)_T \neq 0$ for $j = 1, \ldots, m_\mu$.

Proof. Let $\overline{\lambda_j^\dagger}$ denote the complex conjugate of $\lambda_j^\dagger$ for $j = 1, \ldots, m_\mu$. Then it follows from Theorems 6.1, 6.3, 6.10, and arguments similar to those used in the proof of Theorem 6.9 that $\overline{\mu}$ is an eigenvalue of the problem (2.1), that the $\overline{\lambda_j^\dagger}, j = 1, \ldots, m_\mu$, are precisely the eigenvalues of the system (1.1-4) whose second components are $\overline{\mu}$, and that $\overline{u}_j(x) = \psi^*(x, \overline{\lambda_j^\dagger})$ for $j = 1, \ldots, m_\mu$, where $\overline{u}_j(x)$ denotes the complex conjugate of $u_j(x)$.

Now suppose firstly that $(u_j, \overline{u}_j)_T \neq 0$ for $j = 1, \ldots, m_\mu$. Then it follows from Remark 6.10 that there do not exist any generalized eigenfunctions of the system (1.1-4) corresponding to the eigenvalues $\lambda_j^\dagger$ and $\overline{\lambda_j^\dagger}$ for $j = 1, \ldots, m_\mu$, and hence we conclude from Theorem 6.9 that $\dim N_\mu = \dim N_{\overline{\mu}} = m_\mu$ and that $\{u_j\}_1^{m_\mu}$ and $\{\overline{u}_j\}_1^{m_\mu}$ are bases of N_μ and $N_{\overline{\mu}}$, respectively. It follows immediately from Theorem 1.2 that N_μ and $N_{\overline{\mu}}$ form a dual pair of subspaces of the indefinite inner product space V, and hence, in light of Theorem 6.4, μ and $\overline{\mu}$ must be semi-simple eigenvalues of the problem (2.1).

Conversely, if μ is a semi-simple eigenvalue of the problem (2.1) and $\{u_j\}_1^{m_\mu}$ is a basis of M_μ, then it follows from Theorems 1.2 and 6.4 that $(u_j, \overline{u}_j)_T \neq 0$ for $j = 1, \ldots, m_\mu$, and this completes the proof of the theorem.

Theorem 6.15. Suppose that $\mu \neq 0$ is a real eigenvalue of the problem (2.1) such that N_μ is not a definite subspace of the indefinite inner product space V and let $\{\lambda_j^\dagger\}_1^{n_\mu}$

denote the distinct eigenvalues of the system (1.1-4) whose second components are μ. Let $u_j(x) = \psi^*(x, \lambda_j^\dagger)$ for $j = 1, \ldots, n_\mu$. Then μ is a semi-simple eigenvalue of the problem (2.1) if and only if $(u_j, u_j)_T \neq 0$ for $j = 1, \ldots, n_\mu$.

Proof. Theorem 2.6 assures us that $\{u_j\}_1^{n_\mu}$ is a basis of N_μ, while it follows from Theorem 1.2 that N_μ is a non-degenerate subspace of the indefinite inner product space V if and only if $(u_j, u_j)_T \neq 0$ for $j = 1, \ldots, n_\mu$. The assertion of the theorem now follows from Theorem 6.5.

Theorems 2.6, 3.5, 6.9 and 6.14-15 show that if μ is an eigenvalue of the problem (2.1), then other functions, besides eigenfunctions of the system (1.1-4), may have to be introduced in order to construct a basis for M_μ. In particular, if $\mu = 0$ and the inner product $(\ , \ )_T$ is degenerate on N_0, then (see Theorem 3.5) we have had to introduce the associated vectors χ_{jk} in order to obtain a basis for M_0. Thus, since N_0 has a basis consisting of eigenfunctions of (1.1-4), namely the Ψ_j, $j = 1, \ldots, n$, we shall for this reason henceforth refer to the associated vectors χ_{jk} as functions associated with the eigenfunctions Ψ_j. On the other hand, if μ is a non-real eigenvalue of (2.1), then Theorem 6.9 shows that it may not even be possible to construct a basis for N_μ consisting solely of eigenfunctions of the system (1.1-4); indeed generalized eigenfunctions may have to be introduced in order to achieve this end. Referring to Remark 6.4 and Theorems 6.4, 6.9, it is also clear that the y_{j0} of Theorem 6.4 are linear combinations of the eigenfunctions and generalized eigenfunctions of the system (1.1-4) corresponding to the eigenvalues $\{\lambda_j^\dagger\}_1^{m_\mu}$ cited in Theorem 6.9, and thus, when μ is not semi-simple, we shall for this reason henceforth refer to the vectors y_{jk} of Theorem 6.4 for which $p(j) > 1$ and $k \geq 1$ as functions associated with the eigenfunctions and generalized eigenfunctions of (1.1-4) corresponding to the $\{\lambda_j^\dagger\}_1^{m_\mu}$. Finally, if μ is a non-zero real eigenvalue of the problem (2.1) such that N_μ is not a definite subspace of the indefinite inner product space V, then it is clear from Theorems 2.6 and 6.5 that the y_{j0} of this latter theorem are linear combinations of the eigenfunctions of the system (1.1-4) which correspond to those n_μ distinct eigenvalues whose second components are μ. Thus, when μ is not semi-simple, we shall for this reason henceforth refer to the y_{jk} of Theorem 6.5 for which $p(j) > 1$ and $k \geq 1$ as functions associated with the eigenfunctions of (1.1-4) corresponding to those eigenvalues whose second components are μ.

Referring again to Theorems 6.7-8 for terminology, we have next

Theorem 6.16. Suppose that the system (1.1-4) satisfies the condition $(C_{8\ell-1})$ if

$N(T)\cap V$ is degenerate with respect to the inner product $B(\ ,\ )$. Then eigenfunctions and generalized eigenfunctions (if any) of the system (1.1-4) together with their associated functions (if any) are complete in $L^2(\Omega^\#)$ and in $L^2(\Omega^\#; |\omega(x)|dx)$.

Proof. The assertions of the theorem follow immediately from Theorems 6.7-8.

In order to prove the final result of this section, we shall require

Lemma 6.7. If λ is an eigenvalue of the problem (2.1), then $R(A-\lambda T)$ is closed in $\mathcal{H}$.

Proof. Clearly we need only restrict ourselves to the case $\lambda \neq 0$. Then we know from the proof of Theorem 6.1 and [**98**, Chapter 1] that the Pontrjagin space V_0 admits a decomposition into the direct sum of closed subspaces M_λ and $\mathcal{L}_\lambda$, that K_0 is reduced by this decomposition, and that $\lambda^{-1} \in \rho(K_0|\mathcal{L}_\lambda)$. Moreover, if we let $\overline{\mathcal{L}}_\lambda$ denote the closure of $\mathcal{L}_\lambda$ in $\mathcal{H}_0$, then it is a simple matter to deduce from Theorem 3.6 and the statements just preceding Theorem 3.7 that $\mathcal{H}_0$ is the vector sum of the subspaces M_λ and $\overline{\mathcal{L}}_\lambda$. We also know from these last references that A_0^{-1} is a bounded linear transformation of $R(A_0)$ into $\mathcal{H}_0$, and hence if we let $H^2_\#(\Omega)$ denote the closed subspace $H^2(\Omega)\cap\mathcal{H}_0$ of $H^2(\Omega)$, then it follows from Theorem 2.4 and the closed graph theorem (see [**4**, Lemma 13.4, p.210]) that A_0^{-1} is likewise a bounded linear transformation of $R(A_0)$ into $H^2_\#(\Omega)$.

If $S=(A-\lambda T)|\overline{\mathcal{L}}_\lambda$, then it is clear that S is a closed linear transformation from $\overline{\mathcal{L}}_\lambda$ into $R(A_0)$, and we assert furthermore that $R(S)$ is a closed subspace of $R(A_0)$. Indeed, if this is not the case, then we know from [**120**, Theorems 5.10-11, p.233] that there is a sequence $\{u_j\}$ in $D(S)$ with $\|u_j\|=1$ and $Su_j \to 0$ which contains no convergent subsequence. In view of Proposition 6.2 and the results of the preceding paragraph, this implies that $(I-\lambda K_0)u_j \to 0$ in V_0 (with respect to the strong topology), and since $u_j = v_j + w_j$, where $v_j \in M_\lambda$ and $w_j \in \mathcal{L}_\lambda$, we conclude that $(I-\lambda K_0)v_j \to 0$ and $(I-\lambda K_0)w_j \to 0$ in V_0. It follows immediately that in $\mathcal{H}_0, w_j \to 0$ and $(I-\lambda K)v_j \to 0$ (see the beginning of §6.4). Thus, since K is compact and $\{v_j\}$ is a bounded sequence in $\mathcal{H}_0$, we conclude that $\{v_j\}$ has a convergent subsequence, and hence we arrive at the contradiction that $\{u_j\}$ has a convergent subsequence.

Lastly, it is clear from what was said at the beginning of the proof and Theorem 3.6 that

$$R(A-\lambda T) = (A-\lambda T)M_0 \dot{+} (A-\lambda T)M_\lambda \dot{+} (A-\lambda T)(D(A)\cap\mathcal{L}_\lambda), \tag{6.26}$$

where the first space on the right side of (6.26) does not appear if $0 \in \rho(A)$ and is finite dimensional otherwise (see Theorem 3.4-5), while the second space is finite dimensional (see Theorem 6.1). Hence it follows that $\dim[R(A - \lambda T)/R(S)] < \infty$, and since we already know that the range of S is closed in $\mathcal{H}$, we conclude that this is also the case for $R(A - \lambda T)$, which completes the proof of the lemma.

Finally, let $\{\lambda(j)\}_1^\infty$ be an enumeration of the eigenvalues of the system (1.1-4) and let $v_j(x) = \psi^*(x, \lambda(j))$ for $j \geq 1$.

Theorem 6.17. Suppose that the system (1.1-4) satisfies the condition $(C_{8\ell-1})$ if $N(T) \cap V$ is degenerate with respect to the inner product $B(\ ,\ )$. Then in order that the eigenfunctions of the system (1.1-4) be complete in $L^2(\Omega^\#)$ (resp. in $L^2(\Omega^\#; |\omega(x)|dx)$) it is necessary and sufficient that $(v_j, \overline{v}_j)_T \neq 0$ for $j \geq 1$.

Proof. Suppose firstly that $(v_j, \overline{v}_j)_T \neq 0$ for $j \geq 1$. Then it follows from Theorems 2.6, 3.2, 3.4, 6.1, and 6.14-15 that each eigenvalue of the problem (2.1) is semi-simple and that the corresponding eigenspace has a basis consisting solely of eigenfunctions of the system (1.1-4). In light of Theorem 6.7-8, we conclude that the eigenfunctions of the system (1.1-4) are complete in the spaces cited.

Conversely, suppose that the eigenfunctions of (1.1-4) are complete in either of the spaces cited in the theorem. Then we may argue as we did in the proof of Theorem 5.5 to show that if $\mu = 0$ is an eigenvalue of the problem (2.1), then $(\Psi_j, \Psi_j)_T \neq 0$ for $j = 1, \ldots, n$ (see Theorem 3.2). Assume next that μ is a non-real eigenvalue of the problem (2.1), and with the notation of Theorem 6.9, suppose that for some $j, 1 \leq j \leq m_\mu, \int_0^1 A_r(x_r)\phi_r^2(x_r, \lambda_j^\dagger)dx_r = 0$ for $r = 1, 2$. Then referring to the last paragraph of the proof of Theorem 6.9 for terminology, we see from Remark 6.10 that this implies that there exist generalized eigenfunctions of the system (1.1-4) corresponding to the eigenvalue $\lambda_j^\dagger$, namely the $R_{jk}, k = 1, \ldots, p_j$. On the other hand it follows from Theorem 1.2 and the completeness of the eigenfunctions of (1.1-4) in the space concerned that $(R_{j,p_j}, \overline{R}_{j0})_T = 0$. In view of Theorems 2.6, 6.1, 6.3, 6.9, and Remark 6.11, we conclude that $(R_{j,p_j}, v_k)_T = 0$ for $k \geq 1$, and hence $(R_{j,p_j}, R_{j,p_j})_T = 0$. This shows that $R_{j,p_j}(x) = 0$ almost everywhere in $\Omega^\#$, and since R_{j,p_j} is an eigenvector of (2.1) corresponding to the eigenvalue μ, it follows from Proposition 3.1 that R_{j,p_j} vanishes identically, which of course it not possible. Thus we have shown that there are no generalized eigenfunctions of the system (1.1-4) corresponding to the eigenvalue $\lambda_j^\dagger$ for $j = 1, \ldots, m_\mu$, and hence (see Theorem 6.9) $m_\mu = n_\mu$ and $\{\psi^*(x, \lambda_j^\dagger)\}_1^{m_\mu}$ is a basis of

N_μ. Furthermore, if μ is not a semi-simple eigenvalue of the problem (2.1), then it follows from Theorem 6.14 that there exists a $j, 1 \leq j \leq m_\mu$, such that $(u_j, \overline{u}_j)_T = 0$ (here we use the notation of Theorem 6.14). Since $A - \overline{\mu}T$ is the adjoint of $A - \mu T$ in $\mathcal{H}$ (see the penultimate paragraph of §2.3), we conclude from Theorem 6.3, Lemma 6.7, and Remark 6.11 that there exists a $\chi \in D(A)$ such that $(A - \mu T)\chi = Tu_j$. Moreover, referring again to Theorem 6.14 for terminology, it is clear that we may choose χ so that $(\chi, \overline{u}_k)_T = 0$ for every $k, 1 \leq k \leq m_\mu$, for which $(u_k, \overline{u}_k)_T \neq 0$, if such values of k exist. Thus $\chi \in M_\mu$, while it follows from Theorem 1.2 and the fact that the eigenfunctions of the system (1.1-4) are complete in the space concerned that $(\chi, \overline{u}_k)_T = 0$ for $k = 1, \ldots, m_\mu$. We conclude from this last result and Theorems 6.1, 6.3 that $(\chi, v_k)_T = 0$ for $k \geq 1$, and hence it follows that $(\chi, \chi)_T = 0$. This shows that $\chi(x) = 0$ almost everywhere in $\Omega^\#$, and from this fact it is a simple matter to deduce that $u_j(x) = 0$ for $x \in \Omega^\#$. Hence, in light of Proposition 3.1, we arrive at the contradiction that u_j vanishes identically. Thus μ must be a semi-simple eigenvalue of the problem (2.1), and so we conclude from Theorem 6.14 that $(u_j, \overline{u}_j)_T \neq 0$ for $j = 1, \ldots, m_\mu$.

Finally, if $\mu \neq 0$ is a real eigenvalue of the problem (2.1) such that N_μ is not a definite subspace of the indefinite inner product space V, then we may argue as we did in the previous paragraph to deduce that $(u_j, u_j)_T \neq 0$ for $j = 1, \ldots, n_\mu$, where here we employ the terminology of Theorem 6.15. Thus in light of Theorems 2.6 and 6.1, the proof of the theorem is complete.

6.6. The principal subspaces of the problem (2.1)

Theorems 3.5 and 6.4-5 give us a complete picture of the structure of the principal subspaces of the problem (2.1) corresponding to non-semi-simple eigenvalues, if such eigenvalues exist. Moreover, if $\lambda = 0$ is such an eigenvalue, then we have shown in Chapter 3 how the corresponding eigenvectors and associated vectors are related to the system (1.1-4) (see Remark 3.1). It also follows from Theorems 2.6 and 6.9 that if λ is an eigenvalue of (2.1) of the kind considered in either Theorem 6.4 or Theorem 6.5, then the corresponding eigenvectors y_{j0} are finite linear combinations of the eigenfunctions and generalized eigenfunctions of the system (1.1-4), and hence are members of the class U of functions defined in the statements immediately preceding Theorem 3.5. Thus each y_{j0} is a finite linear combination of decomposable tensors of $\mathcal{H} = \mathcal{H}^1 \otimes \mathcal{H}^2$ (see Remark 3.1 for terminology), and we have also shown in Theorems 2.6 and 6.9 how the components of these tensors are related to the system (1.1-4). It remains only to show for the case where λ is not semi-simple, how the remaining vectors y_{jk} of Theorems 6.4-5, the associated vectors, are related to the system (1.1-4). This problem will be the subject of investigation in this section.

Let us firstly fix our attention upon the case where $\mu \neq 0$ is a real eigenvalue of the problem (2.1) such that N_μ is not a definite subspace of the indefinite inner product space V, let $\lambda_j^\dagger = (\nu_j, \mu)$, $j = 1, \ldots, n_\mu$, denote the distinct eigenvalues of the system (1.1-4) whose second components are μ (see Theorems 2.6, 6.5, and 6.15), and put $u_j(x) = \psi^\star(x, \lambda_j^\dagger)$ for $j = 1, \ldots, n_\mu$. Suppose also that μ is not a semi-simple eigenvalue of (2.1). Then assuming that the terms of the sequence $\{u_j\}_1^{n_\mu}$ have been suitably rearranged and relabelled if necessary, it follows from Theorem 6.15 that there exists the integer $m_\mu, 1 \leq m_\mu \leq n_\mu$, such that $(u_j, u_j)_T = 0$ for $j = 1, \ldots, m_\mu$ and $(u_j, u_j)_T \neq 0$ for $j = (m_\mu + 1), \ldots, n_\mu$ if $m_\mu < n_\mu$. Hence referring to the statements preceding Theorem 3.5 for terminology, we now have

Lemma 6.8. If $1 \leq j \leq m_\mu$, then the system (1.1-4) uniquely determines the sequence of functions on I^2, $\{u_{jk}\}_{k=1}^{p_j}$, where $p_j \leq 2\kappa$, having the following properties. For each $k, u_{jk} \in U$, $u_{jk} \in D(A)$, and $(A - \mu T)u_{jk} = Tu_{j,k-1}$, where $u_{j0} = u_j$. Furthermore, $(u_{jk}, u_r)_T = 0$ for $r = 1, \ldots, n_\mu, r \neq j$, $(u_{jk}, u_j)_T = 0$ if $k < p_j$, and $(u_{j,p_j}, u_j)_T \neq 0$.

Proof. Let $g_{0r}(x_r) = \phi_r(x_r, \lambda_j^\dagger)$ for $r = 1, 2$ (recall the terminology of §1.3). Then we observe that the condition $(u_j, u_j)_T = 0$ uniquely determines the constant e_0 such that (3.1), with e_{0r} replaced by e_0, is valid, and hence if we define the $\omega_r(x_r)$ according to (3.2), then $\int_0^1 \omega_r g_{0r}^2 dx_r = 0$ for $r = 1, 2$.

For $r = 1, 2$, let g_{1r} denote the solution of the initial value problem

$$L_r y_r = (-1)^{r-1} h_r(x_r),\ 0 \leq x_r \leq 1,\ y_r(0) = y_r'(0) = 0, \tag{6.27}$$

where $h_r(x_r) = \omega_{0r}(x_r) = \omega_r(x_r) g_{0r}(x_r)$, $' = d/dx_r$, and L_r denotes the differential expression on the left-hand side of (1.2r-1) with $\lambda_1 = \nu_j$ and $\lambda_2 = \mu$. Then we may argue as in the proof of Lemma 3.1 to show that g_{1r} satisfies the boundary condition (1.2r) and that if $f_1(x) = \sum_{r=1}^2 g_{1r}(x_r) g_{0s}(x_s), s = 3 - r$, then $f_1 \in U$, $f_1 \in D(A), (A - \mu T) f_1 = T f_0$, where $f_0 = u_j$, and $(f_1, u_k)_T = 0$ if $k \neq j$. If $(f_1, u_j)_T \neq 0$, then putting $u_{j1} = f_1, p_j = 1$, the proof of the lemma is complete. On the other hand, if $(f_1, u_j)_T = 0$, then (see the proof of Lemma 3.1) this condition uniquely determines the constant e_1 such that (3.8), with e_{1r} replaced by e_1, is valid, and hence if we define the w_{1r} according to (3.9), then $\int_0^1 \omega_{1r} g_{0r} dx_r = 0$ for $r = 1, 2$.

Let us assume that $(f_1, u_j)_T = 0$. Then we may continue with the above arguments to arrive at the following situation. For $i = 1, \ldots, \ell$ and $r = 1, 2$, there exist the unique numbers e_i and functions g_{ir} such that: (1) g_{ir} is the solution of (6.27) with $h_r = \omega_{i-1,r}$, where w_{ir} is given by (3.10), (2) ω_{ir} satisfies (3.11), (3) g_{ir} satisfies the boundary condition (1.2r), and (4) f_i, defined by (3.12), belongs to U, $f_i \in D(A)$, $(A - \mu T) f_i = T f_{i-1}$, and $(f_i, u_k)_T = 0$ for $k = 0, \ldots, n_\mu$. For $r = 1, 2$, let $g_{\ell+1,r}$ denote the solution of (6.27) with $h_r = \omega_{\ell r}$. Then we may argue as we did in the proof of Lemma 3.1 to show that $g_{\ell+1,r}$ satisfies the boundary condition (1.2r) and if $f_{\ell+1}$ is defined by (3.12) with $i = \ell + 1$, then $f_{\ell+1} \in U$, $f_{\ell+1} \in D(A)$, $(A - \mu T) f_{\ell+1} = T f_\ell$, and $(f_{\ell+1}, u_k)_T = 0$ if $k \neq j$. Thus if $(f_{\ell+1}, u_j)_T \neq 0$, then the proof of the lemma is completed by taking $p_j = \ell + 1$ and $u_{jk} = f_k$ for $k = 1, \ldots, p_j$ (the bound for p_j asserted in the lemma will be established below). On the other hand, if $(f_{\ell+1}, u_j)_T = 0$, then we can show as in the proof of Lemma 3.1 that this condition uniquely determines the constant $e_{\ell+1}$ such that (3.13), with $e_{\ell+1,r}$ replaced by $e_{\ell+1}$, is valid, and hence if we define $\omega_{\ell+1,r}$ by taking $i = \ell + 1$ in (3.10), then (3.11) holds with $i = \ell + 1$ for $r = 1, 2$. Thus if $(f_{\ell+1}, u_j)_T = 0$, then we can continue with the above arguments to obtain $f_{\ell+2}$.

Finally, let us show that the above method must terminate after a finite number of steps, so that in the previous paragraph only the alternative $(f_{\ell+1}, u_j)_T \neq 0$ can hold and that this occurs for $\ell \leq 2\kappa - 1$. Indeed, it is clear that the $f_j, j = 0, \ldots, (\ell + 1)$, of the previous paragraph are contained in M_μ, and hence in V_0, by virtue of Theorem 3.8. Moreover, since $(A - \mu T) f_j = T f_{j-1}$ for $j = 0, \ldots, (\ell + 1)$, where $f_{-1} = 0$, we see that if in the first paragraph of the proof of Theorem 6.1 we take $\lambda = \mu$, $p = \ell + 2$, and define the $z_k, 0 \leq k \leq \ell + 1$, according to (6.2) with u_k replaced by f_k, then we obtain $(K_0 - \mu^{-1} I) z_j = z_{j-1}$ for $j = 0, \ldots, (\ell + 1)$, where $z_{-1} = 0$. In other words, μ^{-1} is an eigenvalue of the compact selfadjoint operator K_0 acting in the Pontrjagin space V_0 and $\{z_j\}_0^{\ell+1}$ is a corresponding Jordan chain. In virture of [**47**, Theorem 4.9, p.191] we see that we must have $\ell + 2 \leq 2\kappa + 1$, and this completes the proof of the lemma.

Theorem 6.18. Suppose that $\mu \neq 0$ is a real eigenvalue of the problem (2.1) such that N_μ is not a definite subspace of the indefinite inner product space V. Suppose also that μ is not semi-simple and let $n_\mu, m_\mu, \{\lambda_j^\dagger\}_1^{n_\mu}$, and $\{u_j\}_1^{n_\mu}$ be defined as above. Then the system (1.1-4) uniquely determines the sequence of functions $u_{jk}, j = 1, \ldots, m_\mu, k = 1, \ldots, p_j$, where $2\kappa \geq p_1 \geq \ldots \geq p_{m_\mu} \geq 1$, such that $u_{jk} \in U$, $u_{jk} \in D(A)$, and $(A - \mu T) u_{jk} = T u_{j,k-1}$, where $u_{j0} = u_j$. Furthermore, $(u_{jk}, u_r)_T = 0$ for $r = 1, \ldots, n_\mu, r \neq j$, $(u_{j,k}, u_j)_T = 0$ if $k < p_j$, and $(u_{j,p_j}, u_j)_T \neq 0$. Finally, the u_{jk} together with the u_j form a linearly independent set in $\mathcal{H}$ and $M_\mu = span\{u_j\} \dot{+} span\{u_{jk}\}$.

Proof. For $j = 1, \ldots, m_\mu$, let u_{jk} and p_j be defined according to Lemma 6.8. Then without loss of generality we can suppose that $p_1 \geq \ldots \geq p_{m_\mu} \geq 1$ since this can always be achieved if necessary by rearranging the $\lambda_j^\dagger$ and relabelling them suitably. Hence all the assertions of the theorem, except the final two, now follow from Lemma 6.8. On the other hand, by appealing to Proposition 3.1 and Lemma 6.2 and by arguing as in the proof of Theorem 3.5, it is a simple matter to verify that the u_{jk} together with the u_j form a linearly independent set in $\mathcal{H}$, and moreover, if we put $M = span\{u_j\} \dot{+} span\{u_{jk}\}$, then we can also show that $\mathcal{H} = M \dot{+} (TM)^\perp$. By arguing again as in the proof of Theorem 3.5, we readily deduce that $M = M_\mu$, and this completes the proof of the theorem.

Remark 6.12. We shall at times in the sequel appeal to Theorem 6.5. Consequently, the importance of Theorem 6.18 lies in the fact that it shows us that when the eigenvalue λ of Theorem 6.5 is not semi-simple, then each associated vector y_{jk} defined in that theorem is a member of U, and hence is a finite linear combination of decomposable tensors of $\mathcal{H}^1 \otimes \mathcal{H}^2$ (see Remark 3.1). Furthermore, we have shown in the proof of Lemma 6.8 just how the components of each of these latter tensors are related to the system (1.1-4).

For the remainder of this section we shall suppose that μ is a non-real eigenvalue of the problem (2.1). Then turning our attention again to Theorem 6.9, we have shown in the last paragraph of its proof that N_μ has a basis $\{R_{jk}\}, j = 1, \ldots, m_\mu, k = 0, \ldots, p_j$. For our purposes it will be more convenient to replace the basis $\{R_{jk}\}$ by the basis $\{h_{jk}\}$ defined in the following way. If $1 \leq j \leq m_\mu$ and $p_j = 0$, then let $h_{j0} = R_{j0}$. Now suppose that for some $j, 1 \leq j \leq m_\mu$, $p_j > 0$. Then we know from the proof of Theorem 6.9 that : (1) the eigenfunction $\theta_0(x_1) = \theta(x_1, \nu_j)$ of (6.17),(1.2) for the eigenvalue ν_j has multiplicity p and $\theta_k(x_1) = (1/k!)(D^k\theta)(x_1, \nu_j)$, $k = 1, \ldots, (p-1)$, is a system of functions associated with θ_0, (2) the eigenfunction $\varsigma_0(x_2) = \varsigma(x_2, \nu_j)$ of (6.19),(1.4) for the eigenvalue ν_j has multiplicity $p^\star$ and $\varsigma_k(x_2) = (1/k!)(D^k\varsigma)(x_2, \nu_j)$, $j = 1, \ldots, (p^\star - 1)$, is a system of functions associated with ς_0, (3) $p_j = \min\{p-1, p^\star - 1\}$, and (4) $R_{jk}(x) = \sum_{r=0}^k \theta_{k-r}(x_1)\varsigma_r(x_2)$ for $k = 0, \ldots, p_j$. We have also shown in the proof of Theorem 6.9 that $I_{rk} = (\theta_r, \overline{\theta}_k)_1$ is constant for $r + k =$ constant and $I_{rk} = 0$ if $r + k < p - 1$, $I_{rk} \neq 0$ if $r + k = p - 1$. Hence it follows that there exist the constants $\{c_r^\dagger\}_0^{p-1}$, depending upon j and with $c_0^\dagger \neq 0$, such that if we put $y_{jk} = \sum_{r=0}^k c_{k-r}^\dagger \theta_r$ for $k = 0, \ldots, (p-1)$, then $(y_{jr}, \overline{y}_{jk})_1$ is 0 or 1 according to whether $r + k \neq p - 1$ or $r + k = p - 1$. We note from Lemmas 6.3-5 and (6.21) that the eigenfunction y_{j0} of the problem (6.17),(1.2) for the eigenvalue ν_j has multiplicity p and that $\{y_{jk}\}_{k=1}^{p-1}$ is a system of functions associated with y_{j0}. If $\mathcal{H}_2$ denotes the Hilbert space $L^2(0 < x_2 < 1; A_2(x_2)dx_2)$ and $(\ ,\ )_2$ denotes the inner product in $\mathcal{H}_2$, then similarly we can show that $J_{rk} = (\varsigma_r, \overline{\varsigma}_k)_2$ is constant for $r + k =$ constant and $J_{rk} = 0$ if $r + k < p^\star - 1$, $J_{rk} \neq 0$ if $r + k = p^\star - 1$. Hence there exist the constants $\{d_r^\dagger\}_0^{p^\star-1}$, depending upon j and with $d_0^\dagger \neq 0$, such that if we put $z_{jk} = \sum_{r=0}^{k} d_{k-r}^\dagger \varsigma_r$ for $r = 0, \ldots, (p^\star - 1)$, then $(z_{jr}, \overline{z}_{jk})_2$ is 0 or 1 according to whether $r + k \neq p^\star - 1$ or $r + k = p^\star - 1$. Moreover, the eigenfunction z_{j0} of the problem (6.19),(1.4) for the eigenvalue ν_j has multiplicity $p^\star$ and $\{z_{jk}\}_{k=1}^{p^\star-1}$ is a system of functions associated with z_{j0}. We now define the h_{jk} for $k = 0, \ldots, p_j$ by putting $h_{jk}(x) = \sum_{\ell=0}^{k} y_{j,k-\ell}(x_1) z_{j\ell}(x_2)$. It is clear from Lemma 6.6 that the $h_{jk}, j = 1, \ldots, m_\mu$, $k = 0, \ldots, p_j$ is a basis of N_μ.

Theorem 6.19. Suppose that μ is a non-real eigenvalue of the problem (2.1) and that $\dim N_\mu = n_\mu \in \mathbb{N}$. Let $\lambda_j^\dagger = (\nu_j, \mu)$, $j = 1, \ldots, m_\mu \leq n_\mu$, denote the distinct eigenvalues of the system (1.1-4) whose second components are μ. Let the integers p_j and functions $h_{jk}(x), j = 1, \ldots, m_\mu, k = 0, \ldots, p_j$, be defined as above. Then μ is a semi-simple eigenvalue of the problem (2.1) if and only if

$$\prod_{j=1}^{m_\mu} \det\big((h_{jk}, \overline{h}_{jr})_T\big)_{k,r=0}^{p_j} \neq 0. \tag{6.28}$$

Proof. To begin with, we note from Lemma 6.7 and the penultimate paragraph of §2.3 that $R(A - \mu T)$ is closed and the adjoint of $A - \mu T$ is precisely $A - \overline{\mu} T$. Now suppose that μ is not a semi-simple eigenvalue of the problem (2.1). Then there is a $u \neq 0$ in N_μ and a $v \in M_\mu$ such that $(A - \mu T)v = Tu$. In view of Theorem 6.3 this implies that $(u, \overline{h}_{jk})_T = 0$ for $j = 1, \ldots, m_\mu, k = 0, \ldots, p_j$. Hence since the h_{jk} form a basis for N_μ, it follows from Remark 6.11 that (6.28) cannot hold.

Conversely, suppose that (6.28) is false. Then there is a $u \neq 0$ in N_μ such that $(u, \overline{h}_{jk})_T = 0$ for $j = 1, \ldots, m_\mu, k = 0, \ldots, p_j$. Thus $Tu \in R(A - \mu T)$, and hence there is a $v \in D(A)$ such that $(A - \mu T)v = Tu$. This shows that μ is not a semi-simple eigenvalue of the problem (2.1) and completes the proof of the theorem.

Continuing with our above discussion, we shall assume from now on that μ is not a semi-simple eigenvalue of the problem (2.1). Hence, in view of Theorem 6.19, we see that by rearranging and relabelling the terms involved we can suppose without loss of generality that : (i) there exists the integer s_1, $1 \leq s_1 \leq m_\mu$, such that $\delta_j = \det\big((h_{jk}, \overline{h}_{jr})_T\big)_{k,r=0}^{p_j} = 0$ for $1 \leq j \leq s_1$ and $\delta_j \neq 0$ for $s_1 < j \leq m_\mu$ if $s_1 < m_\mu$, and (2) there exists the integer s_2, $0 \leq s_2 \leq s_1$ such that $p_j > 0$ for $1 \leq j \leq s_2$ if $s_2 > 0$

and $p_j = 0$ for $s_2 < j \leq s_1$ if $s_2 < s_1$. We note from Remarks 6.10-11 that if $s_2 < s_1$ and $s_2 < j \leq s_1$, then

$$(h_{j0}, \overline{h}_{ks})_T = 0 \text{ for } k = 1, \ldots, m_\mu, s = 0, \ldots, p_k,$$

and (6.29)

$$\sum_{r=1}^{2} \left| \int_0^1 A_r(x_r)\phi_r^2(x_r, \lambda_j^\dagger)dx_r \right| \neq 0.$$

Lemma 6.9. Suppose that $s_2 < s_1$ and $s_2 < j \leq s_1$. Then the system (1.1-4) uniquely determines the sequence of functions on I^2, $\{u_{jk}\}_{k=1}^{p_j^\dagger}$, where $p_j^\dagger \leq \kappa - 1$, having the following properties. For each $k, u_{jk} \in U, u_{jk} \in D(A)$, and $(A - \mu T)u_{jk} = Tu_{j,k-1}$, where $u_{j0} = h_{j0}$. Moreover, $(u_{jk}, \overline{h}_{ri})_T = 0$ for $1 \leq r \leq m_\mu, r \neq j, 0 \leq i \leq p_r, (u_{jk}, \overline{h}_{j0})_T = 0$ if $k < p_j^\dagger$, and $(u_{j,p_j^\dagger}, \overline{h}_{j,0})_T \neq 0$.

Proof. Let $g_{01}(x_1) = y_{j0}$ and $g_{02}(x_2) = z_{j0}(x_2)$. Then as a consequence of (6.29) there exists the unique constant e_0 such that (3.1), with e_{0r} replaced by e_0, is valid, and hence if we define the $\omega_r(x_r)$ according to (3.2), then $\int_0^1 \omega_r g_{0r}^2 dx_r = 0$ for $r = 1, 2$. For $r = 1, 2$, let L_r, $h_r(x_r), \omega_{0r}(x_r)$, and $g_{1,r}(x_r)$ be defined as in the proof of Lemma 6.8. Then we may argue as in the proof of Lemma 3.1 to show that g_{1r} satisfies the boundary condition (1.2r), and if $f_1(x) = \sum_{r=1}^{2} g_{1r}(x_r)g_{0s}(x_s)$, $s = 3 - r$, then $f_1 \in U, f_1 \in D(A), (A - \mu T)f_1 = Tf_0$, where $f_0 = h_{j0}$, and $(f_1, \overline{h}_{k0})_T = 0$ if $k \neq j$. We are now going to show that if $k \neq j$, then

$$(f_1, \overline{h}_{k\ell})_T = 0 \text{ for } \ell = 0, \ldots, p_k. \tag{6.30}$$

It is clear that we need only prove (6.30) for the case $0 < \ell \leq p_k$, and in proving this result we shall make use of the fact that $(y_{j0}, \overline{y}_{k\ell})_1 = (z_{j0}, \overline{z}_{k\ell})_2 = 0$ for $\ell = 0, \ldots, p_k$ (the assertion for the space $\mathcal{H}_1$ is an immediate consequence of the results given in the proof of Theorem 6.9 and the assertion for the space $\mathcal{H}_2$ is similarly proved). Moreover, we shall also make use of the fact, which follows from arguments similar to those used in the proof of Theorem 1.2, that if $1 \leq r \leq 2$, if $L_r Y_r = (-1)^{r-1}F_r$ almost everywhere in $0 \leq x_r \leq 1$, if Y_r satisfies the boundary condition (1.2r), and if $\phi_{i1}(x_1) = y_{ki}(x_1), \phi_{i2}(x_2) = z_{ki}(x_2)$, then

$$(\nu_j - \nu_k)\int_0^1 A_r Y_r \phi_{ir} dx_r = \int_0^1 F_r \phi_{ir} dx_r + \int_0^1 A_r Y_r \phi_{i-1,r} dx_r, \tag{6.31}$$

where $\phi_{-1,r} = 0$. Thus if $0 < \ell \leq p_k$ and $s = 3 - r$, then it follows from (3.7) that

$$\begin{aligned}(f_1, \overline{h}_{k\ell})_T &= \sum_{i=0}^{\ell}\sum_{r=1}^{2}(-1)^{r-1}\Big[\int_0^1 A_r g_{1r}\phi_{\ell-i,r} dx_r \int_0^1 \omega_s g_{0s}\phi_{is} dx_s + \\ &+ \int_0^1 A_r g_{0r}\phi_{\ell-i,r} dx_r \int_0^1 \omega_s g_{1s}\phi_{is} dx_s\Big] \\ &= (\nu_j - \nu_k)^{-1}\sum_{i=o}^{\ell}\sum_{r=1}^{2}(-1)^{r-1}\Big(\int_0^1 \omega_r g_{0r}\phi_{\ell-i,r} dx_r + \\ &+ \int_0^1 A_r g_{1r}\phi_{\ell-i-1,r} dx_r\Big)\int_0^1 \omega_s g_{0s}\phi_{is} dx_s \\ &= (\nu_j - \nu_k)^{-1}\sum_{i=0}^{\ell-1}\sum_{r=1}^{2}(-1)^{r-1}\int_0^1 A_r g_{1r}\phi_{\ell-1-i,r} dx_r \int_0^1 \omega_s g_{0s}\phi_{is} dx_s \\ &= (\nu_j - \nu_k)^{-1}(f_1, \overline{h}_{k,\ell-1})_T,\end{aligned}$$

and hence (6.30) follows immediately. It is also clear from the definitions of the terms involved and Proposition 3.1 that f_1 and f_0 are linearly independent vectors of M_μ, while we know from the proof of Theorem 6.1 that M_μ is a neutral subspace of the Pontrjagin space V_0. Hence it follows from [**47**, Lemma 1.1, p.184] that $\kappa \geq 2$. Thus if $(f_1, \overline{h}_{j0})_T \neq 0$, then putting $u_{j1} = f_1$ and $p_j^\dagger = 1$, the proof of the lemma is complete. If $(f_1, \overline{h}_{j0})_T = 0$, then we can argue as in the proof of Lemma 3.1 to show that this condition uniquely determines the constant e_1 such that (3.8), with e_{1r} replaced by e_1, is valid, and if the ω_{1r} are defined according to (3.9), then $\int_0^1 \omega_{1r} g_{0r} dx_r = 0$ for $r = 1, 2$.

Let us assume that $(f_1, \overline{h}_{j0})_T = 0$. Then we can continue with the above arguments to arrive at the following situation. For $i = 1, \ldots, \ell$ and $r = 1, 2$, there exist the unique numbers e_i and functions g_{ir} such that : (1) g_{ir} is the solution of (6.27) with $h_r = \omega_{i-1,r}$, where ω_{ir} is given by (3.10), (2) ω_{ir} satisfies (3.11), (3) g_{ir} satisfies the boundary condition (1.2r), and (4) f_i, defined by (3.12), belongs to U, $f_i \in D(A)$, $(A - \mu T)f_i = Tf_{i-1}$, and $(f_i, \overline{h}_{kp})_T = 0$ for $k = 1, \ldots, m_\mu$ and $p = 0, \ldots, p_k$. For $r = 1, 2$, let $g_{\ell+1,r}$ denote the solution of (6.27) with $h_r = \omega_{\ell r}$. Then we can show that $g_{\ell+1,r}$ satisfies the boundary condition (1.2r), and if $f_{\ell+1}$ is defined by (3.12) with $i = \ell + 1$, then $f_{\ell+1} \in U$, $f_{\ell+1} \in D(A)$, $(A - \mu T)f_{\ell+1} = Tf_\ell$, and $(f_{\ell+1}, \overline{h}_{k0})_T = 0$ for $k \neq j$. We are now going to show that if $k \neq j$, then

$$(f_{\ell+1}, \overline{h}_{kp})_T = 0 \text{ for } p = 0, \ldots, p_k. \tag{6.32}$$

Clearly we need only consider the case $0 < p \leq p_k$, and for such a p we have on putting $s = 3 - r$ and defining the ϕ_{ir} as above,

$$(\nu_j - \nu_k)(f_{\ell+1}, \overline{h}_{kp})_T =$$

$$= \sum_{i=0}^{p} \sum_{r=1}^{2} (-1)^{r-1} \sum_{q=0}^{\ell} \int_0^1 \omega_{qr} \phi_{p-i,r} dx_r \int_0^1 \omega_s g_{\ell-q,s} \phi_{is} dx_s +$$

$$+ \sum_{i=0}^{p-1} \sum_{r=1}^{2} (-1)^{r-1} \sum_{q=0}^{\ell+1} \int_0^1 A_r g_{qr} \phi_{p-1-i,r} dx_r \int_0^1 \omega_s g_{\ell+1-q,s} \phi_{is} dx_s$$

$$= \Sigma_1 + \Sigma_2,$$

where we have made use of (3.7) and (6.31). It is clear that $\Sigma_2 = (f_{\ell+1}, \overline{h}_{k,p-1})_T$, while arguments similar to those used in the proof of Lemma 3.1 show that $\Sigma_1 = -\sum_{i=1}^{\ell} e_i (f_{\ell-i}, \overline{h}_{kp})_T$. Thus it follows that (6.32) is valid. Consequently, if $(f_{\ell+1}, \overline{h}_{j0})_T \neq 0$, then the proof of the lemma is completed by taking $p_j^\dagger = \ell + 1$ and $u_{ji} = f_i$ for $i = 1, \ldots, p_j^\dagger$ (the bound for $p_j^\dagger$ asserted in the lemma will be established below). On the other hand, if $(f_{\ell+1}, \overline{h}_{j0})_T = 0$, then we can show that this condition uniquely determines the constant $e_{\ell+1}$ such that (3.13), with $e_{\ell+1,r}$ replaced by $e_{\ell+1}$, is valid, and hence, as in the proof of Lemma 6.8, we can continue with the above method to obtain $f_{\ell+2}$.

Finally, we assert that the above method must terminate after a finite number of steps, so that in the previous paragraph only the alternative $(f_{\ell+1}, \overline{h}_{j0})_T \neq 0$ can hold and this occurs for $\ell \leq \kappa - 2$. Indeed, we can argue as in the proof of Lemma 3.1 to show that the vectors $\{f_i\}_0^{\ell+1}$ above form a linearly independet set in M_μ, and hence, for reasons already explained, we must have $\ell + 2 \leq \kappa$. This proves the assertion and completes the proof of the lemma.

Suppose next that $0 < s_2$ and $1 \leq j \leq s_2$. Then referring to the discussion just preceding Theorem 6.19, we see that there is no loss of generality in supposing from now on that $p^\star = p + p^\#$, where $p^\# \geq 0$, and hence $p_j = p - 1$. Now let $g_{01}^{(k)} = y_{jk}$ for $k = 0, \ldots, p_j, g_{02}^{(k)} = z_{jk}$ for $k = 0, \ldots, (p_j + p^\#)$, $e_{0k} = \sum_{i=0}^{k} \int_0^1 B_1 g_{01}^{(k-i)} g_{01}^{(i)} dx_1$ for $k = 0, \ldots, p_j$, and $e'_{0k} = \sum_{i=0}^{k} \int_0^1 B_2 g_{02}^{(k-i)} g_{02}^{(i)} dx_2$ for $k = 0, \ldots, (p_j + p^\#)$. Then if we fix our attention upon the matrix $B^{(0)} = (b_{ik}^{(0)})_{i,k=0}^{p_j}$, where $b_{ik}^{(0)} = (h_{ji}, \bar{h}_{jk})$, it follows from the definitions of the terms involved that $b_{ik}^{(0)}$ is constant for $i+k =$ constant, and

$$\begin{aligned} b_{ik}^{(0)} &= 0 \text{ if } i+k < p_j \\ &= e'_{0,i+k-p_j} - e_{0,i+k-p_j-p\#} \text{ if } p_j \le i+k \le 2p_j, \end{aligned} \tag{6.33}$$

where we are to take $e_{0r} = 0$ if $r < 0$. By hypothesis there exists the integer σ_1, $0 \le \sigma_1 \le p_j$, such that $b_{ik}^0 = 0$ for $i+k \le p_j + \sigma_1$ and $\det(b_{ik}^{(0)})_{i,k=\sigma_1+1}^{p_j} \ne 0$ if $\sigma_1 < p_j$. Hence, in light of Remark 6.11 and the discussion given in the first paragraph of the proof of Theorem 6.19, it follows that $Th_{jk} \in R(A-\mu T)$ for $k = 0, \ldots, \sigma_1$ and $Th_{jk} \notin R(A-\mu T)$ if $\sigma_1 < k \le p_j$. We thus conclude (see §2.3) that there are functions associated with the eigenvectors h_{jk} of (2.1) if and only if $0 \le k \le \sigma_1$.

It would appear that at this stage we could argue with (6.33) as we argued with (6.29) in the proof of Lemma 6.9 for the case $s_2 < j \le s_1$, to obtain an analogous result for the case $1 \le j \le s_2$. However, these arguments, when applied to the present situation, lead after several stages to equations of such complexity that we are not able to arrive at a general result by an inductive argument. On the other hand, these equations do suggest what the final result should be, and accordingly we are led to make the following

Conjecture 6.1. Suppose that $s_2 > 0$ and $1 \le j \le s_2$. Then the system (1.1-4) uniquely determines the sequence of integers $\{\sigma_k\}_1^{p_j^\dagger}$ (where $\sigma_k = \sigma_k(j)$, i.e., the σ_k depend upon j) and the sequence of functions on I^2, $\{u_{jk}^{(i)}\}$, $k = 1, \ldots, p_j^\dagger$, $i = 0, \ldots, \sigma_k$, with the following properties : (1) $p_j = \sigma_0 \ge \sigma_1 \ge \ldots \ge \sigma_{p_j^\dagger} \ge 0$, (2) $\sum_{k=0}^{p_j^\dagger}(1+\sigma_k) \le \kappa$, (3) for each i, k, $u_{jk}^{(i)} \in U$, $u_{jk}^{(i)} \in D(A)$, and $(A-\mu T)u_{jk}^{(i)} = Tu_{j,k-1}^{(i)}$, where $u_{j0}^{(i)} = h_{ji}$, (4) $(u_{jk}^{(i)}, \bar{h}_{r\ell})_T = 0$ if $0 \le k < p_j^\dagger$, $0 \le i \le \sigma_{k+1}$, $1 \le r \le m_\mu$, $0 \le \ell \le p_r$, (5) $(u_{jk}^{(i)}, \bar{h}_{r\ell})_T = 0$ if $0 \le k \le p_j^\dagger$, $\sigma_{k+1} < i \le \sigma_k$ (where we take $\sigma_{p_j^\dagger+1} = -1$), $1 \le r \le m_\mu$, $r \ne j$, and $0 \le \ell \le p_r$, (6) $(u_{jk}^{(i)}, \bar{h}_{j\ell})_T = 0$ if $0 \le k < p_j^\dagger$, $\sigma_{k+1} < i \le \sigma_k$, and $0 \le \ell \le \sigma_{k+1}$, and (7) $\det((u_{jk}^{(i)}, \bar{h}_{j\ell})_T)_{i,\ell=\sigma_{k+1}+1}^{\sigma_k} \ne 0$ if $0 \le k \le p_j^\dagger$ and $\sigma_{k+1} < \sigma_k$. Moreover,

$$u_{jk}^{(i)} = \sum_{r=0}^{k}\sum_{\ell=0}^{i} g_{r1}^{(\ell)} g_{k-r,2}^{(i-\ell)} + \sum_{r=0}^{k-1}\sum_{s=0}^{k-1}\sum_{\ell=0}^{\sigma_r}\sum_{q=0}^{\sigma_s} c_{ijk}^{\ell qrs} g_{r1}^{(\ell)} g_{s2}^{(q)}$$

for $1 \le k \le p_j^\dagger$, $0 \le i \le \sigma_k$, where the $c_{ijk}^{\ell qrs}$ are constants uniquely determined by the system (1.1-4) and the $g_{sr}^{(\ell)}$, $s = 1, \ldots, p_j^\dagger$, $\ell = 0, \ldots, \sigma_s$, $r = 1, 2$, are functions on $0 \le x_r \le 1$ which satisfy the boundary condition (1.2r) and are defined recursively as follows: $g_{sr}^{(\ell)}$ is the solution of (6.27) (with L_r and $'$ defined precisely as in the proof of Lemma 6.8) when

$$h_r = B_r \left[g_{s-1,r}^{(\ell)} + \sum_{m=0}^{s-2} \sum_{q=\sigma_{m+1}+1}^{\sigma_m} d_{\ell s}^{mq} g_{mr}^{(q)} \right] - A_r \left[g_{sr}^{(\ell-1)} + \sum_{m=0}^{s-1} d_{\ell s}^{m} g_{mr}^{(\sigma_m)} \right], \quad (6.34)$$

where $g_{sr}^{(-1)} = 0$, the $d_{\ell s}^{mq}$ and $d_{\ell s}^{m}$ are constants uniquely determined by the system (1.1-4), and the summation $\sum_{m=0}^{s-2}$ in the first bracket on the right side of (6.34) is to be omitted if $s = 1$, while the summation $\sum_{q=\sigma_{m+1}+1}^{\sigma_m}$ is to be omitted if $\sigma_m = \sigma_{m+1}$.

It would appear that the proof of the conjecture requires methods which differ from those used in this book. However, we shall now show that our methods do suffice under a certain restriction and in the process indicate how the conjecture has been arrived at. Accordingly, let us suppose that $0 \le k \le \sigma_1$. Then we have just seen that there do appear vectors associated with h_{jk}. If $\{u_i\}_1^r$ is such a system of vectors (see §2.3), then r is called the length of this associated system, and h_{jk} is said to have multiplicity ℓ if there is a system of vectors associated with h_{jk} of length $\ell - 1$, but no system of length ℓ.

Lemma 6.10. If the multiplicity of h_{j0} does not exceed 3, then Conjecture 6.1 is true.

Proof. For $k = 0, \ldots, \sigma_1$ and $r = 1, 2$, let us define the functions $g_{1r}^{(k)}$ recursively as follows : let $g_{1r}^{(k)}$ denote the solution of (6.27) when

$$h_r = \omega_{0r}^{(k)} = B_r g_{0r}^{(k)} - A_r g_{1r}^{(k-1)} - e_{0k} A_r g_{0r}^{(p_j)},$$

where $g_{1r}^{(-1)} = 0$. By appealing to (6.33) it is not difficult to verify that

$$\int_0^1 \omega_{0r}^{(k)} g_{0r}^{(0)} dx_r = 0 \text{ for } k = 0, \ldots, \sigma_1 \text{ and } r = 1, 2, \quad (6.35)$$

and hence we can now argue as we did in the proof of Lemma 3.1 to show that the $g_{1r}^{(k)}$ satisfy the boundary condition (1.2r). Let

$$f_{1k} = \sum_{r=0}^{1} \sum_{i=0}^{k} g_{r1}^{(i)} g_{1-r,2}^{(k-i)} + \sum_{i=0}^{k} e_{0i} \sum_{\ell=k+1-i}^{p_j} g_{01}^{(\ell)} g_{02}^{(k+p_j+1-i-\ell)} \quad (6.36)$$

for $k = 0, \ldots, \sigma_1$, where in the second expression on the right side of (6.36) the index $i = 0$ is omitted if $k = \sigma_1 = p_j$. Then in light of what has been said above it is a simple matter to verify that for each k, $f_{1k} \in U$, $f_{1k} \in D(A)$, and $(A - \mu T)f_{1k} = Th_{jk}$. Furthermore, familiar arguments involving (6.31), (6.35-36), the orthogonality results concerning the principal vectors of the operator $\mathcal{A}$ defined in the proof of Theorem 6.9, and the analogues of these latter results for the system (6.19), (1.4), show that $(f_{1k}, \overline{h}_{r\ell})_T = 0$ if $0 \le k \le \sigma_1$, $1 \le r \le m_\mu$, $r \ne j$, and $0 \le \ell \le p_r$. Also we can argue as in the proof of Theorem 3.5 to show that the vectors $\{f_{1k}\}_0^{\sigma_1}$ and $\{h_{jk}\}_0^{\sigma_0}$ form a linearly independent set in M_μ, and hence, for reasons already explained, we must have $\sum_{i=0}^{1}(1 + \sigma_i) \le \kappa$. Thus if $\det B^{(1)} \ne 0$, where $B^{(1)}$ denotes the matrix $(b_{ik}^{(1)})_{i,k=0}^{\sigma_1}$ and $b_{ik}^{(1)} = (f_{1i}, \overline{h}_{jk})_T$, then the conjecture follows with $p_j^\dagger = 1$ and $u_{j1}^{(k)} = f_{1k}$ for $k = 0, \ldots, \sigma_1$.

To deal with the remaining case, let us fix our attention upon the matrix $B^{(1)}$. Then lengthy calculations involving the definitions of the terms involved show that $b_{ik}^{(1)}$ is constant for $i + k =$ constant,

$$b_{ik}^{(1)} = 0 \text{ if } i + k < \sigma_1$$

$$\begin{aligned}
&= -\Bigg[\sum_{\ell=0}^{s-\sigma_1} e_{0\ell} \int_0^1 A_s g_{12}^{(\sigma_1)} g_{02}^{(s-\sigma_1-\ell)} dx_2 - \\
&\quad - \sum_{r=0}^{s-p_j} e_{0r} \sum_{\ell=r}^{r+p_j-\sigma_1-1} \int_0^1 B_2 g_{02}^{(\ell+\sigma_1+1-r)} g_{02}^{(s-\sigma_1-\ell)} dx_2 - \\
&\quad - \sum_{r=s+1-p_j}^{s-\sigma_1} e_{0r} \sum_{\ell=r}^{s-\sigma_1} \int_0^1 B_2 g_{02}^{(\ell+\sigma_1+1-r)} g_{02}^{(s-\sigma_1-\ell)} dx_2 - \\
&\quad - \sum_{r=0}^{s-p_j} \int_0^1 B_2 g_{12}^{(s-p_j-r)} g_{02}^{(r)} dx_2 \Bigg] + \\
&\quad + \Bigg[\sum_{\ell=0}^{s-\sigma_1} e_{0,\ell-p^\#} \int_0^1 A_1 g_{11}^{(\sigma_1)} g_{01}^{(s-\sigma_1-\ell)} dx_1 - \\
&\quad - \sum_{r=0}^{s-p_j-p^\#} e_{0r} \sum_{\ell=r}^{r+p_j-\sigma_1-1} \int_0^1 B_1 g_{01}^{(\ell+\sigma_1+1-r)} g_{01}^{(s-\sigma_1-\ell-p^\#)} dx_1 - \\
&\quad - \sum_{r=s+1-p_j-p^\#}^{s-\sigma_1-p^\#} e_{0r} \sum_{\ell=r}^{s-\sigma_1-p^\#} \int_0^1 B_1 g_{01}^{(\ell+\sigma_1+1-r)} g_{01}^{(s-\sigma_1-\ell-p^\#)} dx_1 -
\end{aligned} \tag{6.37}$$

$$- \sum_{r=0}^{s-p_j-p^{\#}} \int_0^1 B_1 g_{11}^{(s-p_j-r-p^{\#})} g_{01}^{(r)} dx_1 \Big]$$

if $i+k=s$ and $\sigma_1 \leq s \leq 2\sigma_1$,

and where the limits of summation are to be interpreted appropriately. Hence if we assume from now on that $\det B^{(1)} = 0$, then there exists the integer σ_2, $0 \leq \sigma_2 \leq \sigma_1$ such that $b_{ik}^{(1)} = 0$ for $i+k \leq \sigma_1 + \sigma_2$ and $\det\big(b_{ik}^{(1)}\big)_{i,k=\sigma_2+1}^{\sigma_1} \neq 0$ if $\sigma_2 < \sigma_1$.

Let us firstly fix our attention upon the case where $\sigma_1 = p_j$. Then it follows from the foregoing results that $Tf_{1k} \in R(A-\mu T)$ for $k = 0, \ldots, \sigma_2$. With this in mind let us now define the functions $g_{2r}^{(k)}, k = 0, \ldots, \sigma_2$, $r = 1, 2$, recursively as follows : let $g_{2r}^{(k)}$ denote the solution of (6.27) when

$$h_r = \omega_{1r}^{(k)} = B_r g_{1r}^{(k)} - A_r g_{2r}^{(k-1)} - e_{0k} A_r g_{1r}^{(p_j)} - e_{1k} A_r g_{0r}^{(p_j)},$$

where $g_{2r}^{(-1)} = 0$ and e_{1k} is determined by the condition that $\int_0^1 \omega_{11}^{(k)} g_{01}^{(0)} dx_1 = 0$. Then by appealing to (6.37) and the definitions of the terms involved we may verify that

$$\int_0^1 \omega_{1r}^{(k)} g_{0r}^{(0)} dx_r = 0 \text{ for } k = 0, \ldots, \sigma_2 \text{ and } r = 1, 2, \tag{6.38}$$

and hence it follows as before that $g_{2r}^{(k)}$ satisfies the boundary condition (1.2r). Consequently, if we let

$$\begin{aligned} f_{2k} = {} & \sum_{r=0}^{2} \sum_{i=0}^{k} g_{r1}^{(i)} g_{2-r,2}^{(k-i)} + \sum_{r=0}^{1} \sum_{i=0}^{k} e_{0i} \sum_{\ell=k+1-i}^{p_j} g_{r1}^{(\ell)} g_{1-r,2}^{(k+p_j+1-i-\ell)} + \\ & + \sum_{i=0}^{k} e_{1i} \sum_{\ell=k+1-i}^{p_j} g_{01}^{(\ell)} g_{02}^{(k+p_j+1-i-\ell)} + \\ & + \sum_{i=0}^{k} e_{0i} \sum_{\ell=k+2-i}^{p_j} e_{0\ell} \sum_{s=k+p_j+2-i-\ell}^{p_j} g_{01}^{(s)} g_{02}^{(k+2p_j+2-i-\ell-s)} \end{aligned} \tag{6.39}$$

for $k = 0, \ldots, \sigma_2$, then direct calculations show that for each k, $f_{2k} \in U$, $f_{2k} \in D(A)$, $(A-\mu T) f_{2k} = T f_{1k}$, and $(f_{2k}, \overline{h}_{r\ell})_T = 0$ if $1 \leq r \leq m_\mu, r \neq j$, and $0 \leq \ell \leq p_r$. Also we can argue as in the proof of Theorem 3.5 to show that the vectors $\{f_{2k}\}_0^{\sigma_2}$, $\{f_{1k}\}_0^{\sigma_1}$, and $\{h_{jk}\}_0^{\sigma_0}$ form a linearly independent set in M_μ, and hence we must have $\sum_{i=0}^{2}(1+\sigma_i) \leq \kappa$. Thus if $\det B^{(2)} \neq 0$, where $B^{(2)}$ denotes the matrix

$$B^{(2)} = \left(b_{ik}^{(2)}\right)_{i,k=0}^{\sigma_2}, \; b_{ik}^{(2)} = \left(f_{2i}, \overline{h}_{jk}\right)_T, \tag{6.40}$$

then the conjecture follows with $p_j^\dagger = 2$ and $u_{jk}^{(i)} = f_{ki}$ for $k = 1, 2$ and $i = 0, \ldots, \sigma_k$. To deal with the remaining case, let us fix our attention upon the matrix $B^{(2)}$. Then lengthy calculations involving (6.33), (6.37-6.39) show that $b_{ik}^{(2)} = 0$ if $i + k < \sigma_2$ and that $b_{ik}^{(2)} =$ constant for $i + k = \sigma_2$. Thus if we henceforth suppose that $\det B^{(2)} = 0$, then it follows that $(f_{20}, \overline{h}_{jk})_T = 0$ for $k = 0, \ldots, \sigma_2$. Putting $f_{2,0}^{\#} = f_{2,0}$ if $\sigma_2 = \sigma_1$ and $f_{20}^{\#} = f_{20} + \sum_{\ell=\sigma_2+1}^{\sigma_1} c_\ell f_{1\ell}$ otherwise, where the constants c_ℓ are chosen so that $(f_{20}^{\#}, \overline{h}_{jk})_T = 0$ for $k = (\sigma_2 + 1), \ldots, \sigma_1$, it follows immediately that $f_{20}^{\#} \in R(A - \mu T)$, and hence we arrive at the contradiction that h_{j0} has multiplicity exceeding 3. Thus $\det B^{(2)} \neq 0$, and this completes the proof of the lemma for the case under consideration.

Suppose next that $\sigma_1 < p_j$ and let

$$f_{1k}^\dagger = f_{1k} + \sum_{\ell=\sigma_1+1}^{p_j} \alpha_{k\ell} h_{j\ell} \quad \text{for} \quad k = 0, \ldots, \sigma_2, \tag{6.41}$$

where the $\alpha_{k\ell}$ are chosen so that

$$(f_{1k}^\dagger, \overline{h}_{j\ell})_T = 0 \quad \text{for} \quad \ell = (\sigma_1 + 1), \ldots, p_j. \tag{6.42}$$

Then it follows from the foregoing results that $T f_{1k}^\dagger \in R(A - \mu T)$ for $k = 0, \ldots, \sigma_2$. Moreover, we observe from (6.37) and the definitions of the terms involved that $e_{0i} = 0$ for $0 \leq i \leq \min\{\rho - 1, \sigma_2\}$, where $\rho = p_j - \sigma_1$. Hence, because of this latter result, it will be convenient for us to treat the two cases, $\sigma_2 < \rho$ and $\sigma_2 \geq \rho$, separately.

Suppose firstly that $\sigma_2 < \rho$. Then for this case the orthogonality conditions (6.42) uniquely determine the constants $\{e_i^{\#}\}_0^{\sigma_2}$ such that in (6.41) we must have

$$\begin{aligned} \alpha_{k\ell} &= e_{k+\sigma_1+1-\ell}^{\#} \quad \text{if} \quad \sigma_1 + 1 \leq \ell \leq \sigma_1 + 1 + k \\ &= 0 \quad \text{if} \quad \sigma_1 + 1 + k < \ell \leq p_j, \end{aligned} \tag{6.43}$$

for $k = 0, \ldots, \sigma_2$. We now define the functions $g_{2r}^{(k)}$, $k = 0, \ldots, \sigma_2$, $r = 1, 2$, recursively as follows : let $g_{2r}^{(k)}$ denote the solution of (6.27) when

$$h_r = \omega_{1r}^{(k)} = B_r \left(g_{1r}^{(k)} + \sum_{i=0}^{k} e_i^{\#} g_{0r}^{(k+\sigma_1+1-i)} \right) -$$

$$- A_r g_{2r}^{(k-1)} - e_k^{\#} A_r g_{1r}^{(\sigma_1)} - e_{1k} A_r g_{0r}^{(p_j)},$$

where $g_{2r}^{(-1)} = 0$ and e_{1k} is determined by the condition that $\int_0^1 \omega_{11}^{(k)} g_{01}^{(0)} dx_1 = 0$. Then as a consequence of the orthogonality conditions (6.42) and the definitions of the terms involved, it is not difficult to verify that (6.38) again holds, and hence it follows that the $g_{2r}^{(k)}$ satisfy the boundary condition (1.2r). Thus if we let

$$\begin{aligned} f_{2k} = & \sum_{r=0}^{2}\sum_{i=0}^{k} g_{r1}^{(i)} g_{2-r,2}^{(k-i)} + \\ & + \sum_{i=0}^{k} e_i^{\#} \sum_{\ell=k+1-i}^{k+\sigma_1+1-i} \left(g_{01}^{(\ell)} g_{12}^{(k+\sigma_1+1-i-\ell)} + g_{11}^{(k+\sigma_1+1-i-\ell)} g_{02}^{(\ell)} \right) + \\ & + \sum_{i=0}^{k} e_{1i} \sum_{\ell=k+1-i}^{p_j} g_{01}^{(\ell)} g_{02}^{(k+p_j+1-i-\ell)} + \\ & + \sum_{i=0}^{k} e_i^{\#} \sum_{\ell=\rho}^{\sigma_1} e_{0\ell} \sum_{s=k+\sigma_1+2-i-\ell}^{p_j} g_{01}^{(s)} g_{02}^{(k+p_j+\sigma_1+2-i-\ell-s)} \end{aligned} \tag{6.44}$$

for $k = 0, \ldots, \sigma_2$, where the last term on the right side of (6.44) does not appear if $\sigma_1 < \rho$, then it is readily verified that for each k, $f_{2k} \in U$, $f_{2k} \in D(A)$, $(A - \mu T) f_{2k} = T f_{1k}^{\dagger}$, and $(f_{2k}, \overline{h}_{r\ell}) = 0$ if $1 \le r \le m_\mu$, $r \ne j$, and $0 \le \ell \le p_r$. Also we can show as before that $\sum_{i=0}^{2}(1 + \sigma_i) \le \kappa$. Thus if $\det B^{(2)} \ne 0$, where $B^{(2)}$ is given by (6.40), then the conjecture follows with $p_j^{\dagger} = 2$, $u_{j2}^{(k)} = f_{2k}$ for $0 \le k \le \sigma_2$, $u_{j1}^{(k)} = f_{1k}^{\dagger}$ for $0 \le k \le \sigma_2$, and $u_{j1}^{(k)} = f_{1k}$ for $\sigma_2 < k \le \sigma_1$ if $\sigma_2 < \sigma_1$. To deal with the remaining case, we again fix our attention upon the matrix $B^{(2)}$. Then we can show as before that $b_{ik}^{(2)} = 0$ if $i + k < \sigma_2$ and that $b_{ik}^{(2)} =$ constant for $i + k = \sigma_2$. Thus if we henceforth suppose that $\det B^{(2)} = 0$, then it follows that $(f_{20}, \overline{h}_{jk})_T = 0$ for $k = 0, \ldots, \sigma_2$. Putting

$$\begin{aligned} f_{20}^{\#} &= f_{20} + \sum_{\ell=\sigma_1+1}^{p_j} d_\ell h_{j\ell} \quad \text{if } \sigma_2 = \sigma_1 \\ &= f_{20} + \sum_{\ell=\sigma_2+1}^{\sigma_1} c_\ell f_{1\ell} + \sum_{\ell=\sigma_1+1}^{p_j} d_\ell h_{j\ell} \quad \text{if } \sigma_2 < \sigma_1, \end{aligned}$$

where the constants c_ℓ, d_ℓ are chosen so that $(f_{20}^{\#}, \overline{h}_{jk})_T = 0$ for $\sigma_2 < k \le p_j$, it follows immediately that $f_{20}^{\#} \in R(A - \mu T)$, and hence we arrive at the contradiction that h_{j0} has multiplicity exceeding 3. Thus det $B^{(2)} \neq 0$, and this completes the proof of the lemma for the case under consideration here.

Suppose finally that $\sigma_2 \geq \rho$. Then for this case the orthogonality conditions (6.42) uniquely determine the constants $\{e_i^{\#}\}_0^{\sigma_2}$ such that in (6.41) we must have $\alpha_{k\ell} = e^{\#}_{k+\sigma_1+1-\ell}$ for $\sigma_1 + 1 \leq \ell \leq p_j$ if $k \geq \rho$, while $\alpha_{k\ell}$ is given by (6.43) if $k < \rho$. Moreover, if we bear in mind the definitions of the terms involved and argue with (6.37) and (6.42), then we can also show that $e_{0i} = e^{\#}_{i-\rho}$ for $\rho \leq i \leq \sigma_2$. We now define the functions $g_{2r}^{(k)}$, $k = 0, \ldots, \sigma_2$, $r = 1, 2$, recursively as follows : let $g_{2r}^{(k)}$ denote the solution of (6.27) when

$$h_r = \omega_{1r}^{(k)} = B_r\left(g_{1r}^{(k)} + \sum_{i=k+1-\rho}^{k} e_i^{\#} g_{0r}^{(k+\sigma_1+1-i)}\right) - \\ - A_r g_{2r}^{(k-1)} - e_k^{\#} A_r g_{1r}^{(\sigma_1)} - e_{1k} A_r g_{0r}^{(p_j)},$$

where $e_i^{\#} = 0$ if $i < 0$, $g_{2r}^{(-1)} = 0$, and e_{1k} is determined by the condition that $\int_0^1 \omega_{11}^{(k)} g_{01}^{(0)} dx_1 = 0$. Then as before we can verify that (6.38) is valid, and hence it follows that the $g_{2r}^{(k)}$ satisfy the boundary condition $(1.2r)$. Consequently, if we let

$$f_{2k} = \sum_{r=0}^{2}\sum_{i=0}^{k} g_{r1}^{(i)} g_{2-r,2}^{(k-i)} + \\ + \sum_{i=0}^{k-\rho} e_i^{\#} \sum_{\ell=k+1-i}^{p_j} \left(g_{01}^{(\ell)} g_{12}^{(k+\sigma_1+1-i-\ell)} + g_{11}^{(k+\sigma_1+1-i-\ell)} g_{02}^{(\ell)}\right) + \\ + \sum_{i=k+1-\rho}^{k} e_i^{\#} \sum_{\ell=k+1-i}^{k+\sigma_1+1-i} \left(g_{01}^{(\ell)} g_{12}^{(k+\sigma_1+1-i-\ell)} + g_{11}^{(k+\sigma_1+1-i-\ell)} g_{02}^{(\ell)}\right) + \\ + \sum_{i=0}^{k} e_{1i} \sum_{\ell=k+1-i}^{p_j} g_{01}^{(\ell)} g_{02}^{(k+p_j+1-i-\ell)} + \\ + \sum_{i=0}^{k-\rho} e_i^{\#} \sum_{\ell=k+2-\rho-i}^{\sigma_1} e_{0\ell} \sum_{s=k+\sigma_1+2-i-\ell}^{p_j} g_{01}^{(s)} g_{02}^{(k+p_j+\sigma_1+2-i-\ell-s)} + \\ + \sum_{i=k+1-\rho}^{k} e_i^{\#} \sum_{\ell=\rho}^{\sigma_1} e_{0\ell} \sum_{s=k+\sigma_1+2-i-\ell}^{p_j} g_{01}^{(s)} g_{02}^{(k+p_j+\sigma_1+2-i-\ell-s)}$$

for $k = 0, \ldots, \sigma_2$, where the limits of summation are to be interpreted appropriately, then direct calculations show that for each k, $f_{2k} \in U$, $f_{2k} \in D(A)$, $(A - \mu T) f_{2k} = T f_{1k}^{\dagger}$, and $(f_{2k}, \bar{h}_{r\ell})_T = 0$ if $1 \le r \le m_\mu$, $r \ne j$, and $0 \le \ell \le p_r$. We can also show as before that $\sum_{i=0}^{2}(1+\sigma_i) \le \kappa$, and hence if $\det B^{(2)} \ne 0$, where $B^{(2)}$ is given by (6.40), then the conjecture follows with $p_j^\dagger = 2$, $u_{j2}^{(k)} = f_{2k}$ for $0 \le k \le \sigma_2$, $u_{j1}^{(k)} = f_{1k}^\dagger$ for $0 \le k \le \sigma_2$, and $u_{j1}^{(k)} = f_{1k}$ for $\sigma_2 < k \le \sigma_1$ if $\sigma_2 < \sigma_1$. On the other hand, if we suppose that $\det B^{(2)} = 0$, then we can argue as we did for the case $\sigma_2 < \rho$ to arrive at a contradiction. Thus $\det B^{(2)} \ne 0$, and this completes the proof of the lemma.

Theorem 6.20. Suppose that μ is a non–real, non–semi–simple eigenvalue of the problem (2.1). Referring to the above arguments for terminology, suppose also that if $s_2 > 0$, $1 \le j \le s_2$, and the multiplicity of h_{j0} exceeds 3, then Conjecture 6.1 is valid. Then the system (1.1-4) uniquely determines the sequence of functions on I^2, $\{u_{jk}^{(i)}\}$, $j = 1, \ldots, s_2$, $k = 1, \ldots, p_j^\dagger$, $i = 0, \ldots, \sigma_k(j)$, if $s_2 > 0$, and the sequence of functions on I^2, $\{u_{jk}\}$, $j = (s_2+1), \ldots, s_1$, $k = 1, \ldots, p_j^\dagger$, if $s_2 < j \le s_1$, such that : (1) if $s_2 > 0$ and $1 \le j \le s_2$, then $p_j = \sigma_0(j) \ge \sigma_1(j) \ge \ldots \ge \sigma_{p_j^\dagger}(j) \ge 0$, (2) $\ell_\mu = \sum_{j=1}^{s_2}\sum_{k=0}^{p_j^\dagger}(1+\sigma_k(j)) + \sum_{j=s_2+1}^{s_1}(1+p_j^\dagger) + \sum_{j=s_1+1}^{m_\mu}(1+p_j) \le \kappa$ if $0 < s_2 < s_1 < m_\mu$, with a similar result holding if the s_i do not satisfy the inequalities shown, (3) if $s_2 > 0$ and $1 \le j \le s_2$, then the $u_{jk}^{(i)}$ have all the properties asserted in Conjecture 6.1, and (4) if $s_2 < j \le s_1$, then the u_{jk} have all the properties asserted in Lemma 6.9. Moreover, the vectors $u_{jk}^{(i)}$ (which only appear if $s_2 > 0$) together with the vectors u_{jk} (which only appear if $s_2 < s_1$) and the vectors h_{jk} form a linearly independent set in $\mathcal{H}$ whose span is precisely M_μ.

Proof. If $s_2 > 0$ and $1 \le j \le s_2$, then let us define the $u_{jk}^{(i)}$ according to Conjecture 6.1, while if $s_2 < s_1$ and $s_2 < j \le s_1$, then let us define the u_{jk} according to Lemma 6.9. Then assertions (1), (3), and (4) of the theorem follow from the foregoing results. To prove the other assertions we shall suppose for simplicity that $0 < s_2 < s_1 < m_\mu$; the remaining cases can be more easily treated. Then we define the vectors v_r, $r = 1, \ldots, \ell_\mu$ as follows : let

$$v_{\ell_\mu(r,k)+i+1} = u_{rk}^{(i)} \text{ for } 1 \le r \le s_2,\ 0 \le k \le p_r^\dagger,\ 0 \le i \le \sigma_k(r),$$

where $\ell_\mu(r,k) = \sum_{j=1}^{r-1}\sum_{\ell=0}^{p_j^\dagger}(1+\sigma_\ell(j)) + \sum_{\ell=k+1}^{p_r^\dagger}(1+\sigma_\ell(r))$ (the first term is to be omitted if $r = 1$ and the second term if $k = p_r^\dagger$),

$$v_{\ell_\mu(r)+p_r^\dagger-k+1} = u_{rk} \text{ for } s_2 < r \le s_1,\ 0 \le k \le p_r^\dagger,$$

where $\ell_\mu(r) = \sum_{j=1}^{s_2}\sum_{\ell=0}^{p_j^\dagger}(1+\sigma_\ell(j)) + \sum_{j=s_2+1}^{r-1}(1+p_j^\dagger)$ (where the second term is to be omitted if $r = s_2 + 1$), and

$$v_{\ell_\mu(r)+k+1} = h_{rk} \quad \text{for} \quad s_1 + 1 \le r \le m_\mu,\ 0 \le k \le p_r,$$

where $\ell_\mu(r) = \sum_{j=1}^{s_2}\sum_{\ell=0}^{p_j^\dagger}(1+\sigma_\ell(j)) + \sum_{j=s_2+1}^{s_1}(1+p_j^\dagger) + \sum_{j=s_1+1}^{r-1}(1+p_j)$ (where the third term is to be omitted if $r = s_1 + 1$). By appealing to Remark 6.11, Lemma 6.9, Conjecture 6.1, and by arguing as in the proof of Theorem 3.5, it is not difficult to verify that

$$\det\left((v_i,\overline{v}_j)_T\right)_{i,j=1}^{\ell_\mu} \neq 0. \tag{6.45}$$

It follows immediately from (6.45) that the v_i form a linearly independent set in $\mathcal{H}$, and hence if $M = span\{v_i\}$, then $\dim M = \ell_\mu$. Since $M \subset M_\mu$, we conclude from Theorems 6.1–3 that $\ell_\mu \le \kappa$, which proves assertion (2) of the theorem. Finally, (6.45) also shows that $\mathcal{H} = M\dot{+}(TM^*)^\perp$, where M^* is composed of those vectors in $\mathcal{H}$ which are precisely the complex conjugates of the vectors of M, and hence we can now argue as we did in the proof of Theorem 3.5 to show that $M_\mu \subset M$, which completes the proof of the theorem.

Remark 6.13. We shall at times in the sequel appeal to Theorem 6.4. Consequently, the importance of Theorem 6.20 (assuming, of course, that the hypotheses made there are valid) lies in the fact that it shows us that when the eigenvalue λ of Theorem 6.4 is not semi–simple, then each associated vector y_{jk} defined in that theorem is a member of U, and hence is a finite linear combination of decomposable tensors of $\mathcal{H}^1 \otimes \mathcal{H}^2$ (see Remark 3.1 for terminology). Furthermore, we have shown in the paragraph immediately preceding Theorem 6.19, in the proof of Lemma 6.9, and in Conjecture 6.1 just how the components of these latter tensors are related to the system (1.1-4).

6.7. The eigenfunction expansion

In this section we shall be concerned with the eigenfunction expansion associated with the system (1.1-4). Accordingly, fixing our attention upon the definitions given in the statements immediately preceding Proposition 6.3, let us suppose that $\mathcal{M}_1 \neq 0$, let E_1 denote the set consisting of those distinct eigenvalues λ of the problem (2.1) for which $Im\lambda > 0$, and for $f \in \mathcal{H}$ let

$$P_1 f = \sum_{\lambda\in E_1}\sum_{j,k\in I_\lambda}\left(c_{jk}^\lambda(f)y_{jk}^\lambda + d_{jk}^\lambda(f)\overline{y_{jk}^\lambda}\right),$$

where the $c^\lambda_{jk}(f)$, $d^\lambda_{jk}(f)$ are given by (6.9) and the remaining terms are defined in the statements just preceding (6.9). Then it is a simple matter to verify that P_1 is the projection mapping $\mathcal{H}$ onto $\mathcal{M}_1$ along $M_0 \dot{+} \mathcal{M}_2 \dot{+} \mathcal{M}_3 \dot{+} \mathcal{H}_1$, while it follows from Theorem 6.2 that $\dim R(P_1) \le 2\kappa$. It is also important to note from the discussion immediately following the proof of Theorem 6.15 and from Remark 6.13 the relationship of the y^λ_{jk} and $\overline{y^\lambda_{jk}}$ to the system (1.1-4). Lastly, if $\mathcal{M}_1 = 0$, then let $P_1 = 0$.

Suppose next that $\mathcal{M}_2 \neq 0$, let E_2 denote the set consisting of those distinct eigenvalues λ of the problem (2.1) for which λ is real, non–zero, and N_λ is not a definite subspace of the indefinite inner product space V, and for $f \in \mathcal{H}$ let

$$P_2 f = \sum_{\lambda \in E_2} \sum_{j,k \in I_\lambda} c^\lambda_{jk}(f) y^\lambda_{jk},$$

where $c^\lambda_{jk}(f)$ is given by (6.10) and the remaining terms are defined in the statements immediately preceding (6.10). Then it is readily verified that P_2 is the projection mapping $\mathcal{H}$ onto $\mathcal{M}_2$ along $M_0 \dot{+} \mathcal{M}_1 \dot{+} \mathcal{M}_3 \dot{+} \mathcal{H}_1$, while it follows from the proof of Theorem 6.1 that $\dim R(P_2) < \infty$. It is also important to note from the discussion immediately following the proof of Theorem 6.15 and from Remark 6.12 just how the y^λ_{jk} are related to the system (1.1-4). Lastly, if $\mathcal{M}_2 = 0$, then let $P_2 = 0$.

Turning finally to the case where $\mathcal{M}_3 \neq 0$, let E_3 denote the set consisting of those distinct eigenvalues λ of (2.1) for which λ is real, non–zero, and N_λ is a negative definite subspace of the Pontrjagin space V_0. Then fixing our attention upon (6.12), let us arrange the eigenvectors u^λ_j, $j = 1, \ldots, n_\lambda$, $\lambda \in E_3$, into some sequence and denote this sequence by $\{u_{-j}\}^r_0$. It is important to observe from the proof of Theorem 6.6 that $r + 1 \le \kappa$.

If $0 \in \rho(A)$, then let $\lambda(j) = (\lambda_1(j), \lambda_2(j))$, $j \ge 1$, denote an arbitrary enumeration of those distinct real eigenvalues of the system (1.1-4) for which $\lambda_2(j) \neq E_2$ if $E_2 \neq 0$, while if $0 \in \sigma(A)$, then let $\lambda(j) = (\lambda_1(j), \lambda_2(j))$, $j \ge n+1$, denote an arbitrary enumeration of those distinct real eigenvalues of the system (1.1-4) for which $\lambda_2(j) \neq 0$ and $\lambda_2(j) \notin E_2$ if $E_2 \neq 0$. If $0 \in \rho(A)$ or if $0 \in \sigma(A)$ and the inner product $(\ ,\ )_T$ is not degenerate on N_0, then let $\psi_j(x)$ be defined by (5.3) for $j \ge 1$, while if $0 \in \sigma(A)$ and the inner product $(\ ,\ )_T$ is degenerate on N_0, then let $\psi_j(x) = \Psi_j(x)$ for $j = 1, \ldots, m$ (see Theorem 3.2) and let $\psi_j(x)$ be defined by (5.3) for $j > m$. Observe from Theorems 1.2 and 3.2 that $(\psi_j, \psi_k)_T = 0$ if $j \neq k$, while $(\psi_j, \psi_j)_T = \rho_j = \pm 1$ for : (1) $j \ge 1$ if $0 \in \rho(A)$ or $0 \in \sigma(A)$ and the inner product $(\ ,\ )_T$ is not degenerate on N_0, and (2) for $j > m$ if $0 \in \sigma(A)$ and the inner product $(\ ,\ )_T$ is degenerate on N_0. It is also clear from Theorem 2.6 that the eigenvectors u_{-j}, $j = 0,\dots,r$, defined above for the case $\mathcal{M}_3 \neq 0$, as well as the characteristic vectors $\{u_j\}_1^\infty$ of the operator K_2 (resp. K_1), introduced in the proof of Theorem 6.7 (resp. Theorem 6.8), can always be chosen so that the sequence $\{u_j\}_{-r}^\infty$ is but a rearrangement of the sequence of eigenfunctions of the system (1.1-4), $\{\psi_j\}_p^\infty$, where $p = 1$ if $0 \in \rho(A)$ and $p = n+1$ otherwise (if $\mathcal{M}_3 = 0$, then we are to replace the sequence $\{u_j\}_{-r}^\infty$ above by the sequence $\{u_j\}_1^\infty$). For the remainder of this chapter it will always be supposed that the u_j have been chosen in this way.

We are now going to state our main results concerning the eigenfunction expansion associated with the system (1.1-4). Accordingly, recalling Definitions 4.2-3 as well as the definition of ρ_j given above, putting $Pf = \sum_{j=1}^2 P_j f$ for $f \in \mathcal{H}$ (see above), and bearing in mind Remark 6.7, we have firstly

Theorem 6.21. Suppose that the system (1.1-4) satisfies the condition (C_1) and that either $0 \in \rho(A)$ or $0 \in \sigma(A)$ and the inner product $(\ ,\ )_T$ is not degenerate on N_0. Suppose also that $N(T) \cap V$ is not degenerate with respect to the inner product $B(\ ,\ )$. Let f satisfy the hypothesis given in Theorem 4.7. Then

$$f(x) = (Pf)(x) + \sum_{j=1}^{\infty} \rho_j (f, \psi_j)_T \psi_j(x) \quad \text{for} \quad x \in I^2, \tag{6.46}$$

where the series on the right side of (6.46) converges regularly, and hence uniformly, on I^2.

Referring to Theorem 3.5 for the definitions of the χ_{jk} and to (3.17) for the definitions of the $c_j(f)$ and $c_{jk}(f)$, we have next

Theorem 6.22. Suppose that the system (1.1-4) satisfies the condition (C_1), that $0 \in \sigma(A)$, and that the inner product $(\ ,\ )_T$ is degenerate on N_0. Suppose also that $N(T) \cap V$ is not degenerate with respect to the inner product $B(\ ,\ )$. Let f satisfy the hypothesis given in Theorem 4.7. Then

$$\begin{aligned} f(x) = \sum_{j=1}^{m} c_j(f)\psi_j(x) + \sum_{j=1}^{m}\sum_{k=1}^{p_j} c_{jk}(f)\chi_{jk}(x) + (Pf)(x) + \\ + \sum_{j=m+1}^{\infty} \rho_j (f, \psi_j)_T \psi_j(x) \quad \text{for} \quad x \in I^2, \end{aligned} \tag{6.47}$$

where the infinite series on the right side of (6.47) converges regularly, and hence uniformly, on I^2.

If ℓ denotes the integer of Theorem 6.8 and if we recall the definition of I^2_ω given in the notation preceding the proof of Theorem 3.2 and that we have written D_r for $\partial/\partial x_r$, then we also have

<u>Theorem 6.23.</u> Suppose that the system (1.1-4) satisfies the condition $(C_{8\ell-1})$ and that either $0 \in \rho(A)$ or $0 \in \sigma(A)$ and the inner product $(\ , \ )_T$ is not degenerate on N_0. Suppose also that $N(T) \cap V$ is degenerate with respect to the inner product $B(\ , \ )$. Let f be a function of class $C^{8\ell,1}$ on I^2 with the following properties : (1) $f(x)$ and its partial derivatives up to and including the $(8\ell-1)$-th order vanish on Γ, (2) $(D_r^{8\ell}f)(x)$ vanishes on $x_r = 0$ (resp. $x_r = 1$) if $\alpha_r = 0$ (resp. $\beta_r = \pi$) for $r = 1, 2$, and (3) $f(x)$ vanishes in a relatively open subset of I^2 containing I^2_ω. Then the conclusions of Theorem 6.21 remain valid.

<u>Theorem 6.24.</u> Suppose that the system (1.1-4) satisfies the condition $(C_{8\ell-1})$, that $0 \in \sigma(A)$, and that the inner product $(\ , \ )_T$ is degenerate on N_0. Suppose also that $N(T) \cap V$ is degenerate with respect to the inner product $B(\ , \ )$. Let f satisfy the hypotheses given in Theorem 6.23. Then the conclusions of Theorem 6.22 remain valid.

We shall only prove Theorems 6.22 and 6.24; the proofs of the remaining theorems are similar.

<u>Proof of Theorem 6.22.</u> We observe from Theorem 2.4 that $f \in D(A)$, and hence if for $x \in I^2$ we let

$$z_0(x) = \sum_{j=1}^{n} c_j(f)\Psi_j(x) + \sum_{j=1}^{m}\sum_{k=1}^{p_j} c_{jk}\chi_{jk}(x) + (Pf)(x) + \\ + \sum_{\lambda \in E_3}\sum_{j=1}^{n_\lambda} c_j^\lambda(f) u_j^\lambda(x), \tag{6.48}$$

where the $c_j^\lambda(f)$ are given by (6.11) and the n_λ and $u_j^\lambda(x)$ are defined in the statements immediately preceding (6.11), and put $w_0(x) = f(x) - z_0(x)$, then it follows from the definitions of the terms involved that $w_0 \in D(A_0) \cap V_1$. Now let $f_1(x) = (\omega(x))^{-1}(Lf)(x)$ for $x \in \Omega^\#$, $f_1(x) = 0$ for $x \in \Omega \backslash \Omega^\#$. In light of Definition 2.1 and Theorem 2.4, it is easy to see that $f_1 \in V$ and $Af = Tf_1$. Moreover, it is clear from Theorems 3.5 and 6.4-5 that there is a $z_1 \in M_0 \dot{+} M$ such that $Az_0 = Tz_1$, and hence if we put $w_1 = f_1 - z_1$, then $Aw_0 = Tw_1$. In light of Proposition 6.4, it is also a simple matter to deduce from this last equation that $w_1 \in V_1$.

Turning now to the proof of Theorem 6.7, it follows from the foregoing results that there is a $g \in \mathcal{N}$ such that

$$w_0 = K_2 g = \sum_{j=1}^{\infty} \mu_j^{-1} \langle g, v_j \rangle v_j = \sum_{j=1}^{\infty} sgn \mu_j (f, u_j)_T u_j \tag{6.49}$$

in $\mathcal{N}$, where we have made use of (4.7). On the other hand, we may appeal to Proposition 6.6 and argue as in the proof of Theorem 5.8 to show that (5.8) is valid, and hence it follows that the series on the right side of (6.49) converges regularly on I^2. Since $f(x) = z_0(x) + w_0(x)$, the proof of the theorem is complete.

Proof of Theorem 6.24. Turning again to the proof of Theorem 6.8, it is not difficult to verify (see [**154**, Theorem 3, p.352]) that the sequence of non-zero eigenvalues of $K_1^{4\ell} = K_1^{2\ell} JS$ is precisely $\{\mu_j^{-4\ell}\}_1^\infty$ and to this sequence of eigenvalues there corresponds the sequence of eigenvectors $\{v_j\}_1^\infty$, where $v_j = |\mu_j|^{-1/2} u_j$. Hence it follows from arguments similar to those of [**154**, Theorem 3, p.409] that if $S^{1/2}$ denotes the positive square root of S, then $S^{1/2} K_1^{2\ell} J S^{1/2}$ is a compact selfadjoint operator in the Hilbert space V_1 (here the inner product is $\langle \ , \ \rangle_J$) whose sequence of non-zero eigenvalues is precisely $\{\mu_j^{-4\ell}\}_1^\infty$ and to this sequence of eigenvalues there correspond the sequence of eigenvectors $\left\{|\mu_j|^\ell S^{1/2} v_j\right\}_1^\infty$ which forms an orthonormal set in V_1.

We are now going to prove that

$$\sum_{j=1}^{\infty} \mu_j^{-6\ell} v_j^2(x) \le c \quad \text{for} \quad x \in I^2, \tag{6.50}$$

where c denotes a positive constant. Accordingly, for $k \in \mathbb{N}$, let $\{d_j\}_1^k$ be arbitrary real constants, and let

$$g = \sum_{j=1}^{k} d_j \mu_j^{4\ell} S^{1/2} v_j, \ v = K_1^{2\ell} J S^{1/2} g = \sum_{j=1}^{k} d_j v_j. \tag{6.51}$$

Then we may argue as we did in the proof of Theorem 4.5 to show that the mapping $A_0^{-1} : R(A_0) \to H^2(\Omega)$ is bounded, and hence it follows from Proposition 6.6 and (6.51) that

$$\|v\|_{2,\Omega} \leq \delta_1 \left\| TK_1^{2\ell-1} JS^{1/2} g \right\| \leq \delta_2 \|g\|_J,$$

where the δ_j are constants not depending upon k or the d_j and $\|h\|_J = \langle h, h\rangle_J^{1/2}$ for $h \in V_1$. Hence we conclude from the Sobolev imbedding theorem that $|v(x)| \leq \delta_3 \|g\|_J$ for $x \in I^2$, where δ_3 is a constant not depending upon x, k, and the d_j. It now follows from (6.51) that

$$\left| \sum_{j=1}^{k} d_j v_j(x) \right| \leq \delta_3 \left[\sum_{j=1}^{k} d_j^2 \mu_j^{6\ell} \right]^{1/2} \quad \text{for } x \in I^2,$$

and hence taking $d_j = \mu_j^{-6\ell} v_j(x)$ for $j = 1, \ldots, k$, we obtain $\sum_{j=1}^{k} \mu_j^{-6\ell} v_j^2(x) \leq \delta_3^2$ for $x \in I^2$. Since k is arbitrary, (6.50) follows immediately.

Next let f be the function given in the theorem and define the functions $\{f_j\}_1^{4\ell}$ in Ω precisely as in the proof of Theorem 6.8. Then again we have $f_j \in D(A)$ and $Af_j = Tf_{j+1}$ for $j = 0, \ldots, (4\ell - 1)$ and $f_{4\ell} \in V$, where $f_0 = f$. Moreover, if z_0 is the function defined by (6.48), then it follows easily from Theorems 3.5, 6.4-5 that there exists the sequence of vectors $\{z_j\}_1^{4\ell}$ in $M_0 \dot{+} M$ such that $Az_j = Tz_{j+1}$ for $j = 0, \ldots, (4\ell - 1)$. Now let $w_j = f_j - z_j$ for $j = 0, \ldots, 4\ell$. Then it is clear from the definitions of the terms involved that $w_0 \in D(A_0) \cap V_1$, $w_j \in D(A)$ for $j = 1, \ldots, (4\ell - 1)$, $w_{4\ell} \in V$, and $Aw_j = Tw_{j+1}$ for $j = 0, \ldots, (4\ell - 1)$. Thus, in light of Theorem 3.5 and Proposition 6.4, we conclude from these results that $w_j \in V_1$ for $j = 0, \ldots, 4\ell$, and hence we must have $w_0 = K_1^{4\ell} w$, where $w = w_{4\ell}$. It now follows from [**154**, Theorem 6, p.420] and (4.7) that

$$\left\| S^{1/2} \left(w_0 - \sum_{j=1}^{k} sgn\mu_j (w_0, u_j)_T u_j \right) \right\|_J \to 0 \text{ as } k \to \infty. \tag{6.52}$$

Since

$$sgn\mu_j (w_0, u_j)_T u_j(x) = \left\langle S^{1/2} w, |\mu_j|^\ell S^{1/2} v_j \right\rangle_J |\mu_j|^{-3\ell} v_j(x) \text{ for } x \in I^2, \tag{6.53}$$

it follows from (2.2), (6.50), and Proposition 6.6 that $\sum_{j=1}^{\infty} sgn\mu_j (w_0, u_j)_T u_j$ is convergent in V_1. Thus we conclude from (6.52) that in V_1,

$$w_0 = \sum_{j=1}^{\infty} sgn\mu_j (w_0, u_j)_T u_j + h, \tag{6.54}$$

where $Sh = 0$. On the other hand we see from [**154**, Theorem 6, p.420] that

$$\left\| S^{1/2}\left(K_1^{2\ell} w - \sum_{j=1}^{k} \mu_j^{-2\ell} \langle w, v_j \rangle v_j \right) \right\|_J \to 0 \text{ as } k \to \infty,$$

while it is clear that

$$K_1^{2\ell}\left(K_1^{2\ell} w - \sum_{j=1}^{k} \mu_j^{-2\ell} \langle w, v_j \rangle v_j \right) \to h \text{ as } k \to \infty,$$

and hence we conclude that

$$\langle Jh, g \rangle_J = \lim_{k \to \infty} \left\langle S\left(K_1^{2\ell} w - \sum_{j=1}^{k} \mu_j^{-2\ell} \langle w, v_j \rangle v_j \right), g \right\rangle_J = 0$$

for any $g \in V_1$. Thus $Jh = 0$, and so $h = 0$. It now follows from Proposition 6.6 and (6.54) that the series $\sum_{j=1}^{\infty} sgn\mu_j (w_0, u_j)_T u_j$ converges to w_0 in $\mathcal{H}$, while (6.50) and (6.53) show that it converges regularly on I^2, and hence the series converges uniformly to w_0 on I^2. Finally, since it is clear from Theorem 3.5 and Proposition 6.4 that $(w_0, u_j)_T = (f, u_j)_T$ for $j \geq 1$, the proof of the theorem is complete.

Remark 6.14. The facts that V_0 is a Pontrjagin space and that $K_0 = K|V_0$ is a compact selfadjoint operator acting in this space clearly do not depend upon the assumption that $\omega(x)$ assumes both positive and negative values in I^2, but only upon the assumption that B_0 is indefinite on V_0 (see §6.1). For this reason some of the results obtained in this chapter are also valid for the case where $\omega(x) \geq 0$ for $x \in I^2$ (see Chapter 4). Indeed, supposing for the moment that our assumption that $\omega(x)$ assumes both positive and negative values on I^2 is replaced by the assumption that $\omega(x) \geq 0$ on I^2, and referring to Proposition 6.3, it follows from the proof of Theorem 4.1 that for this case $\mathcal{M}_1 = \mathcal{M}_2 = 0$, but $\mathcal{M}_3$, and hence M, may or may not be 0, and it is also clear that Propositions 6.3-6 remain valid. Thus, bearing in mind Theorems 3.4 and 3.7, we conclude from arguments similar to those used in the proofs of Theorems 6.7 and 6.21 that if in Theorem 4.7 we replace the assumption that $\gamma \geq 0$ by the assumptions : (1) $\gamma < 0$ and (2) the inner product $B(\ ,\ )$ is not degenerate on $N(T) \cap V$, then the conclusions of this theorem remain valid (see Remark 4.3).

6.8 Comments

Propositions 6.1-2 of §6.3 are taken from [**85**]. Turning now to §6.4 and fixing our attention for the moment upon the situation where L is an ordinary differential operator (so that (2.1) reduces to a weighted Sturm–Liouville problem), the first published results pertaining to the work of this section appear to be those of Haupt [**102, 103**], Hilb [**105**], and Richardson [**140-142**] who prove the existence of infinitely many positive and negative eigenvalues, with the first and last authors also alluding to the possible existence of non-real eigenvalues. For more recent works we refer to Mingarelli [**128-130**], where it is shown that there are infinitely many positive and negative eigenvalues (see Theorem 6.7) and bounds for the number of non–real eigenvalues and for the number of real non–semi–simple eigenvalues, somewhat similar to those asserted in Theorem 6.2, are established. These results are extended in Allegretto and Mingarelli [**6**], while the asymptotic behaviour of the real eigenvalues is investigated in Mingarelli [**131**] and in Atkinson and Mingarelli [**13**]. For further related results concerning both regular and singular problems see Atkinson and Jabon [**12**], Ćurgus and Langer [**59**], Daho and Langer [**60**], Kaper, Kwong and Zettl [**119**], and Mingarelli [**133**]. Finally, for relevant works concerning Birkhoff regular problems see Eberhard, Freiling, and Schneider [**64, 65**] and Freiling and Kaufmann [**96**].

Fixing our attention next upon the case where L is an elliptic partial differential operator, some results pertaining to Theorems 6.1-2 and 6.6 are given in Fleckinger and Mingarelli [**95**] for the Dirichlet and Neumann problems in $\mathbb{R}^k (k \in \mathbb{N})$ involving an operator not necessarily of the second order and under the assumption that the weight function can only vanish on a set of measure zero. However, these results are of a somewhat limited nature and not much detail is given. For the Dirichlet problem in $\mathbb{R}^k$, with $L = -\Delta + q$ (here Δ is the Laplacian and q a real potential) and under the assumption that the weight function can only vanish on a set of measure zero, some results pertaining to Theorems 6.1-2 and 6.6 are given in Fleckinger [**93**], where it is also shown that there are infinitely many positive and negative eigenvalues and the asymptotic behaviour of the distribution functions of these eigenvalues is investigated. However, the proofs are only sketched. This same problem is dealt with in Allegretto and Mingarelli [**6**] without the restriction that the weight function vanishes only on a set of measure zero, and it is shown that there are infinitely many positive and negative eigenvalues and some results pertaining to Theorems 6.1-2 and 6.6 are established. We remark that the proofs of most assertions of §6.4 follow closely the arguments given in [**85**, §4], while some proofs are also to be found in Faierman [**83**] and Faierman and Roach [**90**]. Note that the proof of Theorem 6.4 is essentially a modification of the proof of the decomposition theorem for a pseudo–unitary space in which there acts a symmetric operator (see Mal'cev [**125**, §108]).

For the case of a general pencil $A-\lambda T$, a proof of Theorem 6.6, modified according to Remark 6.6, has been given by Binding and Seddighi [**39**] under conditions described in §4.4. The final assertions of Theorem 6.7 concerning completeness also follow from their work. It is interesting to note that the hitherto mentioned non–degeneracy condition of these authors implies, in our terminology, that at least one of the spaces, $N(K_1)$ or $M(K_1)$, is a non–degenerate subspace of the Pontrjagin space V_1, where $M(K_1)$ denotes the principal subspace of K_1 for the eigenvalue 0. Since we know from Remark 6.8 that the assumption that $N(K_1)$ is not a degenerate subspace of V_1 is equivalent to the assumption that $N(T)\cap V$ is not degenerate with respect to the inner product $B(\ ,\ )$, and since it is a simple matter to show that if $M(K_1)$ is a degenerate subspace of V_1, then so is $N(K_1)$, it would appear that a sharpening of our results could be obtained by inserting between Theorems 6.7 and 6.8 a theorem containing the hypothesis that $M(K_1)$ is not a degenerate subspace of V_1 and having the same conclusions as Theorem 6.7. However this is only illusory, as simple arguments show that $M(K_1) = N(K_1)$. We would also mention that a certain sharpening of Theorem 6.6 has been given by Binding and Browne [**30**] under conditions described in §4.4. Lastly, the final assertion of Theorem 6.2 and an assertion pertaining to Theorem 6.6 are proved in Mingarelli [**132**] for the pencil $A-\lambda T$ under the assumptions that A is an invertible selfadjoint operator with negative lower bound, T is symmetric, $D(A)\cap D(T)$ is dense in $\mathcal{H}$, and that the eigenvalues of the pencil form a proper subset of $\mathbb{C}$.

Turning to §6.5, Theorem 6.9 appears to have been firstly proved by Faierman [**83**] under stronger assumptions on the coefficients of the system (1.1-4) than supposed here and using the method of contour integration (see Titchmarsh [**149**, Chapter I]). The proof given here is essentially based on the ideas sketched out in Faierman and Roach [**90**]. The importance of this theorem lies in the fact that it shows us that at a non–real eigenvalue of the problem (2.1), it may not be possible to construct a basis for the corresponding eigenspace consisting solely of eigenfunctions of the system (1.1-4), and indeed generalized eigenfunctions may have to be introduced in order to achieve this end (compare this theorem with Theorem 2.6). Apart from the assertion concerning the total number of non–real eigenvalues, a proof of Theorem 6.10 appears to have been firstly given in [**81**] and was based upon the ideas sketched out in [**80**]. The proof given here is a modified version of one given in Faierman and Roach [**90**] under a certain restriction. Theorems 6.11-12 and 6.14-15 are also taken from this last reference, while Theorems 6.13 and 6.17 appear to be new. More general multiparameter systems than considered here have been investigated by Binding and Seddighi [**38**] and Binding [**24**] under assumptions which have already been described in §3.5 and which ensure, in our terminology, that $N(T) \cap V$ is not degenerate with respect to the inner product $B(\ ,\ )$. Thus in particular their results contain the assertions concerning completeness given in Theorem 6.16; further results pertaining to Theorems 6.6 and 6.10 are also given in these papers.

Fixing our attention next upon §6.6, we have already mentioned in §3.5 that the structure of the root subspaces of the pencil $A - \lambda T$ which arises from a more general multiparameter eigenvalue problem than considered here has been the subject of investigation in Binding and Seddighi [**38**] and Binding [**24**]. We have also stated that under certain assumptions these two papers together completely describe, in terms of the original multiparameter system, the structure of the principal subspaces of the pencil $A - \lambda T$ corresponding to its real eigenvalues. Thus in particular, under the assumption that T is invertible, Theorem 6.18 is contained in [**24**]. Our results concerning the principal subspaces of the pencil corresponding to its non–real eigenvalues (see Lemmas 6.9-10 and Theorem 6.20) appear to be new; the validity of Conjecture 6.1 for the general case (i.e., not covered by Lemma 6.10) remains an open question.

Finally, turning to the uniform convergence of the eigenfunction expansion dealt with in §6.7, all the results of this section are essentially modifications of the results given in Faierman and Roach [**90**].

APPENDIX A

A right definite two-parameter eigenvalue problem involving complex potentials

A.1. Introduction

Much of the machinery established in the main part of this book can also be used to prove some basic facts concerning the eigenvalues and eigenfunctions of a right definite two–parameter eigenvalue problem involving complex potentials. Accordingly, we shall take up this problem here, and hence in §A.2 below we introduce the two–parameter eigenvalue problem in question. In §A.3 we introduce the associated elliptic boundary value problem, while in §A.4 we establish some results concerning the eigenvalues and principal vectors of the operator induced in a certain Hilbert space by our elliptic operator, the boundary conditions, and the weight function. Our main results are then obtained in §A.5 by examining the relationship between the eigenvalues and principal vectors of the induced operator and the eigenvalues and eigenfunctions of the two–parameter eigenvalue problem. Finally, in §A.6, we conclude the appendix with some comments.

A.2. Preliminaries

In this appendix we will again be concerned with the simultaneous two–parameter system (1.1-4), where now it is supposed that in $0 \leq x_r \leq 1 (r = 1, 2)$, p_r is positive and Lipschitz continuous, A_r and B_r are real–valued and Lipschitz continuous, and q_r is complex–valued and essentially bounded. If we define the eigenvalues and eigenfunctions of the system (1.1-4) as in §1.2, then it is now our objective to establish some of their basic properties under the assumption that

$$\omega(x) = A_1(x_1)B_2(x_2) - A_2(x_2)B_1(x_1) \neq 0 \text{ in } I^2, \tag{A.1}$$

where throughout this appendix we employ the same notation, unless otherwise stated, as in the main part of the book.

The definiteness condition (A.1) has certain implications which we wish now to investigate. Accordingly, let S^1 denote the unit circle in $\mathbb{R}^2$ with centre at the origin, let $u = (u_1, u_2)$ denote the points of S^1, and let $P_r(x_r, u) = (-1)^{r-1}(u_1 A_r(x_r) - u_2 B_r(x_r))$ for $0 \leq x_r \leq 1$, $u \in S^1$, and $r = 1, 2$.

Proposition A.1. There exist elements $u^\star$ and $u^\dagger$ of S^1 such that : (1) $u^\star$ and $u^\dagger$ are orthogonal in $\mathbb{R}^2$, (2) $(-1)^{r-1}P_r(x_r, u^\star) > 0$ in $0 \le x_r \le 1$ for $r = 1, 2$, and (3) $u_1^\star u_2^\dagger - u_2^\star u_1^\dagger = sgn\omega(x)$.

Proof. Let us define the mapping $f(u) = (f_1(u), f_2(u))$ of S^1 into $\mathbb{R}^2$ in the following way. For each u in S^1 and for $r = 1, 2$ let :

(i) $f_r(u)$ denote the infimum of $P_r(x_r, u)$ in $0 \le x_r \le 1$ if $P_r(x_r, u) > 0$ at every point of this interval,

(ii) $f_r(u) = 0$ if $P_r(x_r, u)$ has at least one zero in $0 \le x_r \le 1$,

(iii) $f_r(u)$ denote the supremum of $P_r(x_r, u)$ in $0 \le x_r \le 1$ if $P_r(x_r, u) < 0$ at every point of this interval.

It is not difficult to verify that $f(u)$ is a continuous mapping of S^1 into $\mathbb{R}^2$, $f(u) \ne 0$, and $f(-u) = -f(u)$. From a result of Borsuk [**48**] it follows that the image of S^1 under f meets every ray emanating from the origin in $\mathbb{R}^2$. Thus there exists the point $u^\star = (u_1^\star, u_2^\star)$ of S^1 and the positive number d such that $(-1)^{r-1}f_r(u^\star) > d$ for $r = 1, 2$. Hence if we take $u^\dagger = (-u_2^\star sgn\omega(x), u_1^\star sgn\omega(x))$, then the proof of the proposition is complete.

As a consequence of Proposition A.1 we see that there is no loss of generality in introducing

Assumption A.1. We will henceforth suppose that the A_r and ω are positive for all values of x_1 and x_2 in their respective intervals.

Indeed, it follows immediately from Proposition A.1 that the conditions guaranteed by the assumption can always be achieved if necessary by means of a change of parameters of the form $\lambda_r = \sum_{s=1}^{2} u_{rs}\lambda'_s$, $r = 1, 2$, where (u_{rs}) is an orthogonal matrix.

Referring again to the system (1.1-4), let $q(x) = q_1(x_1)A_2(x_2) + q_2(x_2)A_1(x_1)$ and $q_0(x) = Req(x)$ for $x \in I^2$. Then another consequence of the definiteness condition (A.1), and which will be used in the sequel, is contained in the following

Proposition A.2. Let δ be any positive number. Then we can always arrange, by introducing a shift in the parameter λ_2 if necessary, that $q_0(x) \ge \delta$ for $x \in I^2$.

Proof. In (1.1) and (1.3) let us introduce the transformation $\lambda_1 = \lambda'_1$, $\lambda_1 = \mu + \lambda'_2$, where $\mu \in \mathbb{R}$ is a parameter which will be specified below. Then the coefficient of

y_r in $(1.2r-1)$, $r=1,2$, becomes $(-1)^{r-1}(\lambda_1' A_r(x_r) - \lambda_2' B_r(x_r)) - Q_r(x_r)$, where $Q_r(x_r) = (-1)^{r-1}\mu B_r(x_r) + q_r(x_r)$. Thus if $Q(x) = Q_1(x_1)A_2(x_2) + Q_2(x_2)A_1(x_1)$ for $x \in I^2$, then $Q_0(x) = ReQ(x) = q_0(x) - \mu\omega(x)$. Hence bearing in mind that $\omega(x) > 0$ for $x \in I^2$, it follows that $Q_0(x) \geq \delta$ for $x \in I^2$ if we choose μ to be the infimum of $(q_0(x) - \delta)/\omega(x)$ in I^2.

A.3. The associated elliptic boundary value problem

Assumption A.1 suggests that one way of dealing with the eigenvalue problem (1.1-4) is to proceed in the opposite direction to that of separation of variables, that is to say, to go back to the original elliptic boundary value problem from which (1.1-4) arose, establish some relevant information concerning the eigenvalues and principal vectors of the operator induced in $\mathcal{H}_\omega = L^2(\Omega; \omega(x)dx)$ (recall that Ω denotes the interior of I^2) by our elliptic operator, the boundary conditions, and $\omega(x)$, and then arrive at our main results by demonstrating their relationship to the eigenvalues and eigenfunctions of (1.1-4).

Accordingly, we proceed as in Chapter 2 to arrive at the boundary value problem (2.1), where all terms are defined as before, and again it will be convenient to refer to (2.1) as BVP I, BVP II, or BVP III according to the conditions given in the statements preceding Definition 2.1. Next let

$$B_0^\dagger(u,v) = \sum_{r=1}^{2}(D_r u, a_r D_r v) + (u, q_0 v) \quad \text{for} \quad u,v \in H^1(\Omega),$$

where $q_0(x) = Req(x)$ and $q(x)$ and the $a_r(x)$ are defined in §2.3. Then with V denoting the closed subspace of $H^1(\Omega)$ given in Definition 2.1, let us introduce in V the sesquilinear form $B_0(u,v)$, where $B_0(u,v)$ is defined according to Definition 2.2 when in that definition we replace $B(u,v)$ and $B^\dagger(u,v)$ by $B_0(u,v)$ and $B_0^\dagger(u,v)$, respectively. It is clear that B_0 is symmetric and densely defined in $\mathcal{H}$, while we may argue as we did in Proposition 2.1 to show that B_0 is coercive over V and that $|B_0(u,v)| \leq c\|u\|_{1,\Omega}\|v\|_{1,\Omega}$ for $u,v \in V$, where c denotes a positive constant. Now let δ be an arbitrary positive number. Then we know from Proposition A.2 that by introducing a shift in the parameter λ_2 of (1.1) and (1.3) if necessary, we can always arrange that $q_0(x) \geq \delta$ for $x \in I^2$. Assuming that this is the case, we can then argue as in the proof of Proposition 2.1 to show that

$$B_0^\dagger(u,u) \geq c_1 \sum_{r=1}^{2} \|D_r u\|^2 + \delta\|u\|^2 \quad \text{for} \quad u \in V,$$

where c_1 denotes a positive constant not depending upon δ. Moreover, if we are dealing with BVP II or BVP III and if we let $\Gamma^\dagger$ denote Γ in the former case and Γ' in the latter case, then it follows from (2.3) that

$$\left|\int_{\Gamma^\dagger} \sigma |tru|^2 ds\right| \leq 2^{-1} c_1 \sum_{r=1}^{2} \|D_r u\|^2 + c_2 \|u\|^2 \quad \text{for} \quad u \in V,$$

where c_2 denotes a positive constant not depending upon δ. Hence if we take $\delta = c_1$ if we are dealing with BVP I and $\delta = c_2 + c_1/2$ otherwise, then it follows that

$$B_0(u,u) \geq 2^{-1} c_1 \|u\|_{1,\Omega}^2 \quad \text{for} \quad u \in V. \tag{A.2}$$

Thus we see that by introducing a shift in the parameter λ_2 of (1.1) and (1.3) if necessary, there is no loss of generality in assuming

Assumption A.2. It will henceforth be supposed that (A.2) is valid.

It follows immediately that the form B_0 is closed and that the lower bound of B_0 is positive. Referring to the discussion following the proof of Proposition 2.2 for details, we henceforth denote by A_0 the selfadjoint operator in $\mathcal{H}$ associated with B_0; and we observe from [**120**, Theorem 2.6, p.323 and p.278] that $0 \in \rho(A_0)$. Furthermore, A_0 has compact resolvent, and hence a discrete spectrum. Also if L_0 denotes the elliptic operator

$$-\sum_{r=1}^{2} D_r a_r(x) D_r + q_0(x),$$

then we can argue as we did in the proof of Theorem 2.4 to show that

Theorem A.1. An element u in V belongs to $D(A_0)$ if and only if $u \in H^2(\Omega)$ and satisfies (2.1b) in the sense of trace. Moreover, $A_0 u = L_0 u$ in $\mathcal{H}$ for $u \in D(A_0)$.

Next let us introduce in $\mathcal{H}$ the operator $Q \in \mathcal{L}(\mathcal{H})$ defined by $(Qf)(x) = (q_0(x) - q(x))f(x)$ and the sesquilinear form $B_1(u,v) = -(Qu,v)$. Since B_1 is B_0-bounded with B_0-bound less than 1 [**120**, p.319], we conclude from [**120**, Theorem 1.33, p.320] that $B = B_0 + B_1$ is sectorial and closed, while it is also clear that $ReB = B_0$. Hence if A denotes the m-sectorial operator associated with B [**120**, Theorem 2.1, p.322], then it follows from [**120**, Corollary 2.3, p.323] that $0 \in \rho(A)$, while a simple argument shows that $A = A_0 - Q$. Furthermore, we may argue with A as we argued with A_0 to show that A has compact resolvent, and hence a discrete spectrum. Lastly, we have just seen that the domain of B is V and $D(A) = D(A_0)$, while from Theorem A.1 it follows that

Theorem A.2. An element u in V belongs to $D(A)$ if and only if $u \in H^2(\Omega)$ and satisfies (2.1b) in the sense of trace. Moreover, $Au = Lu$ in $\mathcal{H}$ for $u \in D(A)$.

To terminate the work of this section, let $Q^\star$ denote the adjoint of Q in $\mathcal{H}$, $B_1^\star$ the adjoint form of B_1 in $\mathcal{H}$ [**120**, p.309], and $L^\star$ the elliptic operator

$$-\sum_{r=1}^{2} D_r a_r(x) D_r + \overline{q}(x),$$

where $\overline{q}(x)$ denotes the complex conjugate of $q(x)$. Then we may argue as in the preceding paragraph to show that the form $B_0 + B_1^\star$ is sectorial and closed, and that the m-sectorial operator associated with this form is precisely $A_0 - Q^\star$. Thus $D(A_0 - Q^\star) = D(A_0)$, and hence in light of Theorems A.1 and A.2, [**120**, Theorem 2.5, p.323], and the fact that $B^\star = B_0 + B_1^\star$, where $B^\star$ denotes the adjoint form of B in $\mathcal{H}$, we conclude that if $A^\star$ denotes the adjoint of A in $\mathcal{H}$, then

Theorem A.3. It is the case that $D(A) = D(A^\star) = D(A_0)$ and $A^\star = A_0 - Q^\star$. Moreover, $A^\star u = L^\star u$ in $\mathcal{H}$ for $u \in D(A^\star)$.

A.4. The Hilbert space $\mathcal{H}_\omega$

For our purposes it will be more convenient to work in the Hilbert space $\mathcal{H}_\omega = L^2(\Omega; \omega(x)dx)$ rather than in $\mathcal{H}$, and we let $(\, , \,)_\omega$ and $\| \, \|_\omega$ denote the inner product and norm, respectively, in $\mathcal{H}_\omega$. Note that $\mathcal{H}$ and $\mathcal{H}_\omega$, considered only as vector spaces, are the same and that $\| \, \|$ and $\| \, \|_\omega$ are equivalent norms in this space. Hence in $\mathcal{H}_\omega$ we may introduce the sesquilinear forms B_0 and B, both with domain V, and it is easy to see that B_0 is densely defined, closed, and symmetric, while B is densely defined, closed, and sectorial. Furthermore, if T denotes the bounded selfadjoint operator in $\mathcal{H}_\omega$ (and in $\mathcal{H}$) defined by $(Tf)(x) = \omega(x)f(x)$, then it is clear that the selfadjoint operator in $\mathcal{H}_\omega$ associated with B_0 is precisely $T^{-1}A_0$, while the m-sectorial operator in $\mathcal{H}_\omega$ associated with B is precisely $T^{-1}A$. It is also clear that $0 \in \rho(T^{-1}A_0), 0 \in \rho(T^{-1}A)$, and that the operators $K_0 = (T^{-1}A_0)^{-1} = A_0^{-1}T$ and $A_0^{-1}Q$ are compact selfadjoint and compact, respectively, in $\mathcal{H}_\omega$. Finally, observing that $1 \in \rho(A_0^{-1}Q)$, let S denote the compact operator in $\mathcal{H}_\omega$ defined by $I + S = (I - A_0^{-1}Q)^{-1}$, and where we note that $-1 \in \rho(S)$.

It follows from the above results that $T^{-1}A$ is a m-sectorial operator in $\mathcal{H}_\omega$ with a vertex $\gamma > 0$. Hence the spectrum of $T^{-1}A$ is contained in the sector $|\arg(z-\gamma)| \leq \theta$ for some θ satisfying $0 < \theta < \pi/2$; and we shall henceforth denote this sector by $\Theta_{\gamma,\theta}$. Moreover, since $(T^{-1}A)^{-1} = (I+S)K_0$, we see also that $T^{-1}A$ has compact resolvent. Thus $T^{-1}A$ has a discrete spectrum, and so in particular $R(T^{-1}A - \mu I)$ is a closed subspace of $\mathcal{H}_\omega$ and $\dim N(T^{-1}A - \mu I) < \infty$ for every $\mu \in \mathbb{C}$. Now let ϵ be any positive number satisfying $0 < \epsilon < \pi/2$ and let $G_\epsilon = \{z | z \in \mathbb{C}, |\arg z| < \epsilon\}$.

Theorem A.4. The eigenvalues of $T^{-1}A$ form a denumerably infinite subset of $\mathbb{C}$ having no finite points of accumulation and are all contained in the sector $\Theta_{\gamma,\theta}$. Furthermore, these eigenvalues are each of finite algebraic multiplicity and, except for possibly a finite number of them, lie in the sector G_ϵ. Finally, the principal vectors of $T^{-1}A$ form a complete set in $\mathcal{H}_\omega$.

In order to prove the theorem, we shall need

Definition A.1. Let $\mathcal{H}^\dagger$ be a separable complex Hilbert space and let $H \in \mathcal{L}(\mathcal{H}^\dagger)$ be compact. Then the eigenvalues of the compact non–negative operator $(H^*H)^{1/2}$, arranged in decreasing order and counted according to multiplicity, form a sequence $\{\mu_j\}$, where $\mu_j \to 0$ if the sequence is infinite. We call the μ_j the s-numbers of the operator H and say that H belongs to the class C_p, $0 < p < \infty$, if $\sum_j \mu_j^p < \infty$ (see [**63**, pp.1088–1089], [**98**, p.27 and p.92]).

Proof of Theorem A.4. We have seen above that K_0 is a compact selfadjoint operator in $\mathcal{H}_\omega$ and it is clear that $K_0 \geq 0$. Hence the s-numbers of K_0 coincide with its eigenvalues (counted according to multiplicity). It is also clear that $T^{1/2}A_0^{-1}T^{1/2}$ is a compact selfadjoint operator in $\mathcal{H}$ with non–negative lower bound, and thus its s-numbers coincide with its eigenvalues (counted according to multiplicity). Moreover, since it is a simple matter to show that the eigenvalues of K_0, repeated according to multiplicity, are precisely those of $T^{1/2}A_0^{-1}T^{1/2}$, we conclude that the s-numbers of K_0 coincide with those of $T^{1/2}A_0^{-1}T^{1/2}$. On the other hand, we know from [**3**] and Theorem A.1 that A_0^{-1}, as an operator acting in $\mathcal{H}$, is of class C_p for any p satisfying $1 < p < \infty$, and hence it follows from [**98**, p.27] that $T^{1/2}A_0^{-1}T^{1/2}$ is of class C_p. Thus K_0 is of class C_p, and so we conclude from [**98**, Theorem 8.1, p.257] that if $K = K_0(I+S)$, then the principal vectors of K form a complete set in $\mathcal{H}_\omega$ and all the characteristic values of K, with the possible exception of at most a finite number of them, lie in the sector G_ϵ.

Next if μ is an eigenvalue of $T^{-1}A$, then let us denote by M_μ and N_μ the principal subspace and eigenspace, respectively, of $T^{-1}A$ corresponding to μ. If μ is an eigenvalue of K, then let $\mathcal{M}_\mu$ and $\mathcal{N}_\mu$ denote the principal subspace and eigenspace, respectively, of K corresponding to μ. Then in light of the foregoing results we see that the theorem will be proved if we can show that μ is an eigenvalue of $T^{-1}A$ if and only if μ is a characteristic value of K and that $M_\mu = (I+S)\mathcal{M}_{1/\mu}$. Accordingly, suppose firstly that μ is an eigenvalue of $T^{-1}A$. Then it is clear that $\mu \neq 0$ and that if $0 \neq u \in N_\mu$, then $(I - \mu K)v = 0$, where $v = (I+S)^{-1}u$. Thus μ is a characteristic value of K and $N_\mu \subset (I+S)\mathcal{N}_{1/\mu}$. Now suppose that μ is not semi-simple and that $u \in M_\mu \backslash N_\mu$. Then there exist the vectors $\{u_j\}_0^{p-1}$, where $p > 1, u_{p-1} = u$, and $u_0 \neq 0$, such that $(T^{-1}A - \mu I)u_j = u_{j-1}$, where $u_{-1} = 0$. Hence $(I - \mu K)v_j = Kv_{j-1}$ for $j = 0, \dots, (p-1)$, where $v_j = (I+S)^{-1}u_j$, and so $(I-\mu K)^p v_{p-1} = 0$, $(I - \mu K)^{p-1}v_{p-1} \neq 0$. We conclude from these arguments that if μ is an eigenvalue of $T^{-1}A$, then μ is a characteristic value of K and $M_\mu \subset (I+S)\mathcal{M}_{1/\mu}$.

Conversely, suppose that μ is a characteristic value of K. If $v \neq 0$ and $(I - \mu K)v = 0$, then v, and hence $u = (I+S)v$, belong to $D(A)$, and $(T^{-1}A - \mu I)u = 0$. Thus we conclude that μ is an eigenvalue of $T^{-1}A$ and $(I+S)\mathcal{N}_{1/\mu} \subset N_\mu$. Now assume that $1/\mu$ is not a semi-simple eigenvalue of K and that $v \in \mathcal{M}_{1/\mu} \backslash \mathcal{N}_{1/\mu}$. Then there exist the vectors $\{v_j\}_0^{p-1}$ in $\mathcal{M}_{1/\mu}$, where $p > 1, v_{p-1} = v$, and $v_0 \neq 0$, such that $(K - \mu^{-1}I)v_j = v_{j-1}$, where $v_{-1} = 0$. Thus the v_j, and hence the $u_j = (I+S)v_j$, belong to $D(A)$ and $(T^{-1}A - \mu I)u_0 = 0$, $(T^{-1}A - \mu I)u_j = -\mu^2 \sum_{r=0}^{j-1}(-\mu)^{j-1-r}u_r$ for $j = 1, \dots, (p-1)$. It follows immediately that $u_{p-1} \in D(A^p)$ and $(T^{-1}A - \mu I)^p u_{p-1} = 0, (T^{-1}A - \mu I)^{p-1}u_{p-1} = (-\mu^2)^{p-1}u_0 \neq 0$. We conclude from these arguments that if μ is a characteristic value of K, then μ is an eigenvalue of $T^{-1}A$ and $(I+S)\mathcal{M}_{1/\mu} \subset M_\mu$. This completes the proof of the theorem.

We may also introduce in $\mathcal{H}_\omega$ the sesquilinear form $B^\star$, with domain V, and show as above that $B^\star$ is densely defined, closed, and sectorial, that the m-sectorial operator in $\mathcal{H}_\omega$ associated with $B^\star$ is precisely $T^{-1}A^\star$, and that Theorem A.4 remains valid when $T^{-1}A$ is replaced by $T^{-1}A^\star$. Now for μ an eigenvalue of $T^{-1}A^\star$, let $M_\mu^\star$ and $N_\mu^\star$ denote the principal subspace and eigenspace, respectively, of $T^{-1}A^\star$ corresponding to μ. Then if we let $(T^{-1}A)^\star$ denote the adjoint of $T^{-1}A$ in $\mathcal{H}_\omega$ and recall the definitions of M_μ and N_μ given in the proof of Theorem A.4, we have

Theorem A.5. It is the case that $(T^{-1}A)^\star = T^{-1}A^\star$, and hence $\rho(T^{-1}A^\star)$ and $\sigma(T^{-1}A^\star)$ are respectively the mirror images of $\rho(T^{-1}A)$ and $\sigma(T^{-1}A)$ with respect to the real axis. Moreover, for each eigenvalue μ of $T^{-1}A, M_{\overline{\mu}}^\star$ (resp. $N_{\overline{\mu}}^\star$) is composed of those vectors in $\mathcal{H}_\omega$ which are precisely the complex conjugates of the vectors of M_μ (resp. N_μ). Finally, if μ and ν are any two distinct eigenvalues of $T^{-1}A$, then M_μ and $M_{\overline{\mu}}^\star$ are orthogonal in $\mathcal{H}_\omega$.

Proof. Since the adjoint form of B in $\mathcal{H}_\omega$ is precisely $B^\star$, it follows from [**120**, Theorem 2.5, p.323] that $(T^{-1}A)^\star = T^{-1}A^\star$; and the remaining assertions made in the first sentence of the theorem now follow from [**120**, Theorem 6.22, p.184]. The assertions of the second sentence of the theorem are proved by appealing to Theorems A.2-3 and arguing as in the proof of Theorem 6.3.

To prove the final assertion of the theorem, let P_μ denote the Riesz projector of $T^{-1}A$ corresponding to the eigenvalue μ, that is, $P_\mu = -(1/2\pi i)\int_C (T^{-1}A - zI)^{-1}dz$, where C is a rectifiable, simple closed curve contained in $\rho(T^{-1}A)$ which enclosed the point μ, but no other point of $\sigma(T^{-1}A)$, and is orientated in the usual sense (see [**120**, pp.178 and 180]). Let P_ν be defined analogously. Lastly, let $Q_{\overline{\nu}}$ denote the Riesz projector of $T^{-1}A^\star$ corresponding to the eigenvalue $\overline{\nu}$. Hence observing from [**120**, pp.181 and 184] that $P_\nu P_\mu = 0$ and $P_\nu^\star = Q_{\overline{\nu}}$, we see that if $u \in M_\mu$ and $v \in M_{\overline{\nu}}^\star$, then $(u,v)_\omega = (P_\mu u, Q_{\overline{\nu}} v)_\omega = (P_\mu u, P_\nu^\star v)_\omega = (P_\nu P_\mu u, v)_\omega = 0$, and this completes the proof of the theorem.

A.5. The system (1.1-4)

We are now going to use the results of §A.4 in order to establish our main results concerning the eigenvalues and eigenfunctions of the system (1.1-4). Accordingly, if we recall the definition of $\psi^\star(x,\lambda)$ given in the statements preceding Theorem 1.2, then we have firstly

Theorem A.6. Let $\lambda^\dagger = (\nu,\mu)$ be an eigenvalue of the system (1.1-4). Then μ is an eigenvalue of $T^{-1}A$ and $\psi^\star(x,\lambda^\dagger)$ a corresponding eigenvector.

Proof. The proof is the same as the proof of Theorem 2.5, except now we use Theorem A.2 in place of Theorem 2.4.

Referring next to the discussion following the proof of Lemma 6.2, but now replacing the condition $\mu \in \mathbb{C}\setminus\mathbb{R}$ there by the condition $\mu \in \mathbb{C}$, we define the generalized eigenfunctions of the system (1.1-4) corresponding to the eigenvalue $\lambda^\dagger = (\nu,\mu)$ precisely as in the statements preceding Lemma 6.6. Hence recalling the definition of N_μ given in the proof of Theorem A.4 and bearing in mind Remark 6.9, we then have

Theorem A.7. Suppose that μ is an eigenvalue of $T^{-1}A$ and $\dim N_\mu = n_\mu$. Then there are precisely m_μ distinct eigenvalues of the system (1.1-4) whose second components are μ, where $1 \le m_\mu \le n_\mu$. Moreover, if we denote these eigenvalues by $\{\lambda_j^\dagger\}_1^{m_\mu}$, then $\{\psi^\star(x,\lambda_j^\dagger)\}_1^{m_\mu}$ is a basis of N_μ if $m_\mu = n_\mu$. On the other hand, if $m_\mu < n_\mu$, then for at least one j there exist generalized eigenfunctions of the system (1.1-4) corresponding to the eigenvalue $\lambda_j^\dagger$, and there is a basis of N_μ consisting of $\{\psi^\star(x,\lambda_j^\dagger)\}_1^{m_\mu}$ together with the generalized eigenfunctions of the system (1.1-4) correponding to those eigenvalues $\lambda_j^\dagger$ for which such generalized eigenfunctions exist.

Proof. The proof is almost the same as that of Theorem 6.9 and will not be repeated.

Remark A.1. Referring again to Theorem A.7, suppose that $1 \le j \le m_\mu$. Then it is not difficult to verify (see Remark 6.10) that in order for there to exist generalized eigenfunctions of the system (1.1-4) corresponding to the eigenvalue $\lambda_j^\dagger$, it is necessary and sufficient that $\int_0^1 A_r(x_r)\phi_r^2(x_r,\lambda_j^\dagger)dx_r = 0$ for $r = 1,2$ (see §1.3 for terminology).

Remark A.2. Suppose that in Theorem A.7, $m_\mu > 1$, $j \ne k$, and u (resp. v) denotes either the eigenfunction $\psi^\star(x,\lambda_j^\dagger)$ (resp. $\psi^\star(x,\lambda_k^\dagger)$) or a generalized eigenfunction of (1.1-4) corresponding to the eigenvalue $\lambda_j^\dagger$ (resp. $\lambda_k^\dagger$) if any such functions exist. Then it is not difficult to verify (see Remark 6.11) that $(u,\overline{v})_\omega = 0$ (recall that $\overline{v}(x) = \overline{v(x)}$).

Recalling the defintion of M_μ given in the proof of Theorem A.4, we now have

Theorem A.8. Suppose that μ is an eigenvalue of $T^{-1}A$ and let $\{\lambda_j^\dagger\}_1^{m_\mu}$ denote the distinct eigenvalues of the system (1.1-4) whose second components are μ. Let $u_j(x) = \psi^\star(x,\lambda_j^\dagger)$ for $j = 1,\dots,m_\mu$. Then μ is a semi–simple eigenvalue of $T^{-1}A$ and $\{u_j\}_1^{m_\mu}$ is a basis of M_μ if and only if $(u_j,\overline{u}_j)_\omega \ne 0$ for $j = 1,\dots,m_\mu$.

Proof. Suppose firstly that $(u_j,\overline{u}_j)_\omega \ne 0$ for $j = 1,\dots,m_\mu$. Then it follows from Remark A.1 that there do not exist any generalized eigenfunctions of the system (1.1-4) corresponding to the eigenvalue $\lambda_j^\dagger$ for $j = 1,\dots,m_\mu$, and hence it follows from Theorem A.7 that $\dim N_\nu = m_\mu$ and $\{u_j\}_1^{m_\mu}$ is a basis of N_μ. Let us now assume that μ is not semi–simple. Then there exist vectors $v \ne 0$ in $D(A)$ and $u \ne 0$ in N_μ

such that $(T^{-1}A - \mu I)v = u$. Hence there exist scalars $\{c_j\}_1^{m_\mu}$, not all zero, such that $u = \sum_{j=1}^{m_\mu} c_j u_j$, while Theorem A.5 shows that $(u, \overline{u}_j)_\omega = 0$ for $j = 1, \ldots, m_\mu$. Thus in light of Remark A.2, we arrive at the contradiction that $c_j = 0$ for $j = 1, \ldots, m_\mu$.

Conversely, suppose that μ is a semi–simple eigenvalue of $T^{-1}A$, that $\{u_j\}_1^{m_\mu}$ is a basis of M_μ, and that for some $j, 1 \leq j \leq m_\mu$, $(u_j, \overline{u}_j)_\omega = 0$. Then it follows from Theorem A.5, Remark A.2, and the fact that $R(T^{-1}A - \mu I)$ is a closed subspace of $\mathcal{H}_\omega$ that $u_j \in R(T^{-1}A - \mu I)$, and hence there exists a $v \in D(A)$ such that $(T^{-1}A - \mu I)v = u_j$. Thus $(T^{-1}A - \mu I)^2 v = 0$ and $(T^{-1}A - \mu I)v \neq 0$, which contradicts the fact that μ is semi–simple.

Returning again to Theorem A.7, let us observe that we can construct the basis $\{h_{jk}\}, j = 1, \ldots, m_\mu$, $k = 0, \ldots, p_j$, of N_μ in precisely the manner indicated in the paragraph following Remark 6.12. Now bearing in mind Theorem A.8, let us fix our attention upon the case where $(u_j, \overline{u}_j)_\omega = 0$ for at least one j, where $u_j(x) = \psi^*(x, \lambda_j^\dagger)$. Then it is not difficult to deduce from Theorem A.5, Remark A.2, and arguments similar to those used in the proof of Theorem 6.19 that μ is a semi–simple eigenvalue of $T^{-1}A$ if and only if (6.28) is valid. Hence if $1 \leq j \leq m_\mu, (u_j, \overline{u}_j)_\omega = 0$, and $\sum_{r=1}^{2} \left| \int_0^1 A_r(x_r) \phi_r^2(x_r, \lambda_j^\dagger) dx_r \right| \neq 0$, then it follows from Remark A.1 that μ cannot be semi-simple. Thus we see that in order to construct a basis for M_μ, other vectors, besides the eigenfunctions and generalized eigenfunctions, may have to be introduced.

The foregoing remarks lead to the following theorem, and we refer to §2.2 for terminology.

Theorem A.9. Suppose that μ is an eigenvalue of $T^{-1}A$ and that there are precisely m_μ distinct eigenvalues of the system (1.1-4) whose second components are μ. Then whether μ is semi-simple or not, M_μ may be decomposed into the direct sum of subspaces $\{\mathcal{L}_j\}_1^{n_\mu}, n_\mu \geq m_\mu$, where $\mathcal{L}_j$ has a basis $\{y_{jk}\}, k = 0, \ldots, (p(j) - 1)$, satisfying : (1) $(T^{-1}A - \mu I)y_{jk} = y_{j,k-1}(y_{j,-1} = 0)$, (2) $(y_{jk}, \overline{y}_{r,\ell})_\omega = 0$ if $j \neq r$ and $(y_{jk}, \overline{y}_{jr})_\omega = 1$ or 0 according to whether $k + r = p(j) - 1$ or $k + r \neq p(j) - 1$, (3) $p(1) \geq p(2) \geq \ldots \geq p(n_\mu) \geq 1$, and (4) $y_{jk} \in C^1(I^2)$ and satisfies the boundary conditions (1.2) and (1.4).

Proof. By hypothesis there exists a $p \in \mathbb{N}$ such that $(T^{-1}A - \mu I)^p M_\mu = 0, (T^{-1}A - \mu I)^{p-1} M_\mu \neq 0$, and hence in M_μ we may introduce the bilinear form $[y, z] = \left((T^{-1}A - \mu I)^{p-1} y, \overline{z}\right)_\omega$. Then we may appeal to Theorems A.2-3 and A.5 and argue as in the proof of Theorem 6.4 to show that $[y, z] = [z, y]$ for every y, z in M_μ. Now suppose that $[z, z] = 0$ for every $z \in M_\mu$. Then as in the proof of Theorem 6.4 we can show that this implies that $[y, z] = 0$ for every y, z in M_μ, and hence in light of Theorem A.5 we see that for every $z \in M_\mu, (T^{-1}A - \mu I)^{p-1}z$ is orthogonal in $\mathcal{H}_\omega$ to all the principal vectors of $T^{-1}A^*$. Since these latter vectors form a complete set in $\mathcal{H}_\omega$, we arrive at the contradiction that $(T^{-1}A - \mu I)^{p-1}z = 0$ for every $z \in M_\mu$. Thus there exists a $z \in M_\mu$ such that $[z, z] \neq 0$.

Let $(T^{-1}A - \mu I)z_r = z_{r-1}$ for $r = 0, \ldots, (p-1)$, where $z_{p-1} = z$. Then it is clear that $z_{-1} = 0$, while it is a simple matter to verify that $I_{jk} = (z_j, \overline{z}_k)_\omega$ is constant for $j + k =$ constant, $I_{jk} = 0$ if $j + k < p - 1$, and $I_{jk} \neq 0$ if $j + k = p - 1$. Consequently, we can argue as in the proof of Theorem 6.4 to show that there exist the unique constants $\{c_k\}_0^{p-1}$, with $c_0 \neq 0$, such that if we put $p(1) = p$ and let $y_{1k} = \sum_{\ell=0}^{k} c_{k-\ell} z_\ell$ for $k = 0, \ldots, (p(1) - 1)$, then $(y_{1k}, \overline{y}_{1\ell})_\omega$ is 1 or 0 according to whether $k + \ell = p(1) - 1$ or $k + \ell \neq p(1) - 1$, while $(T^{-1}A - \mu I)y_{1k} = y_{1,k-1}$, where $y_{1,-1} = 0$. If $\mathcal{L}_1 = span\{y_{1k}\}_0^{p(1)-1} \neq M_\mu$, then M_μ can be represented as the direct sum of $\mathcal{L}_1$ and a subspace $\mathcal{L}$, which is invariant under $T^{-1}A$, such that $(u, \overline{v})_\omega = 0$ for $u \in \mathcal{L}_1$ and $v \in \mathcal{L}$. By repeating the above arguments in the subspace $\mathcal{L}$ instead of M_μ, we can prove that $\mathcal{L}$ contains a subspace $\mathcal{L}_2$ having a basis $\{u_{2k}\}_0^{p(2)-1}$, where the y_{2k} have properties analogous to those of the y_{1k} cited above and $p(2) \leq p(1)$. Hence by a repetition of the above arguments if necessary, we establish the existence of the $\mathcal{L}_j$ and the $y_{jk}, j = 1, \ldots, n_\mu, k = 0, \ldots, (p(j) - 1)$, of the theorem satisfying all the assertions with the exception of assertion (4). Since assertion (4) can be proved by using arguments similar to those used in the proof of assertion (3) of Theorem 6.4, the proof of the theorem is complete.

<u>Remark A.3.</u> We will henceforth refer to the y_{jk} of Theorem A.9 for which $p(j) > 1$ and $k \geq 1$ as functions associated with the eigenfunctions and generalized eigenfunctions (if any) of the system (1.1-4) corresponding to those m_μ eigenvalues whose second components are μ.

<u>Remark A.4.</u> Theorem A.9 gives us a complete picture of the structure of the principal subspace M_μ of $T^{-1}A$ corresponding to the eigenvalue μ. However from the point of view of multiparameter spectral theory it is also important to describe just how the vectors y_{jk} of M_μ are related to the system (1.1-4). In this vein we point out that it is a consequence of Theorem A.7 that the y_{j0} are finite linear combinations of eigenfunctions and generalized eigenfunctions of the system (1.1-4), and hence are members of the class U of functions defined in the statements preceding Theorem 3.5 (see the remarks immediately preceding Lemma 6.6). Thus each y_{j0} is a finite linear combination of decomposable tensors of $\mathcal{H}^1 \otimes \mathcal{H}^2$ (see Remark 3.1 for terminology) and we have shown in

the proof of Theorem 6.9 (see the proof of Theorem A.7) just how the components of these tensors are related to the system (1.1-4). Since it is not difficult to verify that Lemma 6.9 (apart from the assertion that $p_j^\dagger \leq \kappa - 1$), Conjecture 6.1 (apart from assertion (2)), Lemma 6.10, Theorem 6.20 (apart from assertion (2) and with the first sentence of the theorem replaced by the supposition that μ is a non-semi-simple eigenvalue of $T^{-1}A$), and Remark 6.13 (with Theorem 6.4 there replaced by Theorem A.9) apply in full force to the problem under consideration here, similar remarks also hold for the associated functions $y_{jk}, 1 \leq j \leq n_\mu, 1 \leq k \leq p(j) - 1, p(j) > 1$, when μ is not semi-simple, provided, of course, that we assume that the hypotheses of Theorem 6.20 are valid.

As a consequence of Theorems A.4, A.6-7, and A.9 we now have

Theorem A.10. The eigenvalues of the system (1.1-4) form a denumerably infinite subset of $\mathbb{C}^2$ having no finite points of accumulation. Moreover, if we denote these eigenvalues by $\lambda(j) = (\lambda_1(j), \lambda_2(j))$ for $j \geq 1$, then the $\lambda_2(j)$ are all contained in the sector $\Theta_{\gamma,\theta}$ and, except for possibly a finite number of them, also lie in the sector G_ϵ. Finally, the eigenfunctions and generalized eigenfunctions (if any) of the system (1.1-4), together with their associated functions (if any), form a complete set in $\mathcal{H}_\omega$.

From Theorem A.8 we also have

Theorem A.11. In order that the eigenfunctions of the system (1.1-4) form a complete set in $\mathcal{H}_\omega$, it is necessary and sufficient that $(v_j, \overline{v}_j)_\omega \neq 0$ for every $j \geq 1$, where $v_j(x) = \psi^*(x, \lambda(j))$.

Remark A.5. This last theorem generalizes the well known result that the eigenfunctions of the system (1.1-4) are complete in $\mathcal{H}_\omega$ if q_1 and q_2 are both real-valued.

A.6. Comments

There is a large literature devoted to the study of boundary value problems involving ordinary differential operators and elliptic partial differential operators with complex potentials; we refer to Edmunds and Evans [**66**], Naimark [**135**], and the references given therein for further information. Fixing our attention now upon multiparameter eigenvalue problems involving complex potentials, we have already mentioned in §3.5 the relevant discussions of Atkinson [**10, 11**], Binding [**24**], and Isaev [**112**] concerning the principal subspaces associated with such problems. However, as far as the problem pertaining to the completeness of the eigenfunctions is concerned, the only relevant work appears to be that of Faierman [**86**], on which this appendix is based, in spite of the fact that this problem has been the subject of much investigation for the case of real–valued potentials (cf. Atkinson [**11**], Browne [**49**], Faierman [**69**], Sleeman [**143**], and Volkmer [**151**]).

Turning to §A.2, Proposition A.1 is taken from Faierman [**72**] and here we used a well known result of Borsuk to establish the proposition. This result of Borsuk has found much use in right definite multiparameter problems, and we refer to Atkinson [**11**, §9.4] for a further application. In §§A.3-4, we made much use of the perturbation theory of selfajoint operators in order to establish our main results, and in particular, Theorem A.4, which is of fundamental importance to our theory, was proved by appealing to a very significant perturbation result of Keldỹs. Finally, fixing our attention upon §A.5, the importance of Theorm A.7 lies in the fact that it shows us that at an eigenvalue of $T^{-1}A$, it may not be possible to construct a basis of the corresponding eigenspace consisting solely of eigenfunctions of the system (1.1-4), and indeed generalized eigenfunctions may have to be introduced in order to achieve this end. The proof of Theorem A.9 is essentially a modification of the arguments used by Mal'cev in [**125**, §108]. The results concerning the principal subspaces of $T^{-1}A$ given in Remark A.4 appear to be new, while the validity of Conjecture 6.1 (omitting, of course, assertion (2)) for the general case (that is, not covered by Lemma 6.10) remains an open question.

References

1. R.A. Adams, *Sobolev Spaces*, Academic, New York, 1975.
2. S. Agmon, The L_p approach to the Dirichlet problem, *Ann. Scuola Norm. Sup. Pisa* **13**(1959), 405–449.
3. S. Agmon, On the eigenfunctions and on the eigenvalues of general elliptic boundary value problems, *Comm. Pure Appl. Math.* **15**(1962), 119–147.
4. S. Agmon, *Lectures on Elliptic Boundary Value Problems*, Van Nostrand, Princeton, N.J., 1965.
5. S. Agmon, A. Douglis, and L. Nirenberg, Estimates near the boundary for solutions of elliptic partial differential equations satisfying general boundary conditions. I, *Comm. Pure Appl. Math.* **12**(1959), 623–727.
6. W. Allegretto and A.B. Mingarelli, Boundary problems of the second order with an indefinite weight function, *J. Reine Angew. Math.* **398**(1989), 1–24.
7. F.M. Arscott, Two–parameter eigenvalue problems in differential equations, *Proc. London Math. Soc.* **14**(1964), 459–470.
8. F.V. Atkinson, *Discrete and Continuous Boundary Problems*, Academic, New York, 1964.
9. F.V. Atkinson, *Multivariate spectral theory: the linked eigenvalue problem for matrices.* Technical Summary Report No. 431, U.S. Army Mathematics Research Center, Madison, Wisconsin, 1964.
10. F.V. Atkinson, Multiparameter spectral theory, *Bull. Amer. Math. Soc.* **74**(1968), 1–27.
11. F.V. Atkinson, *Multiparameter Eigenvalue Problems*, Vol. 1, Academic, New York, 1972.
12. F.V. Atkinson and D. Jabon, Indefinite Sturm–Liouville problems, in *Proc. 1984 Workshop "Spectral Theory of Sturm–Liouville Operators"*, Argonne, IL, 1984, Reprint ANL–84–73, 31–45.
13. F.V. Atkinson and A.B. Mingarelli, Asymptotics of the number of zeros and of the eigenvalues of general weighted Sturm–Liouville problems, *J. Reine Angew. Math.* **375/376**(1987), 380–393.
14. T. Ya. Azizov and I.S. Iokhvidov, *Linear Operators in spaces with an Indefinite Metric*, Wiley, New York, 1989.
15. R. Beals, Indefinite Sturm–Liouville problems and half–range completeness, *J. Differential Equations* **56**(1985), 391–407.
16. Yu. M. Berezanskii, *Expansions in Eigenfunctions of Selfadjoint Operators*, Amer. Math. Soc., Providence, R.I., 1968.
17. P. Binding, On a problem of B.D. Sleeman, *J. Math. Anal. Appl.* **85**(1982), 291–307. Erratum: ibid **90**, 270–271.
18. P. Binding, Left definite multiparameter eigenvalue problems, *Trans. Amer. Math. Soc.* **272**(1982), 475–486.
19. P. Binding, Multiparameter variational principles, *SIAM J. Math. Anal.* **13**(1982), 842–855.

20. P. Binding, Dual variational approaches to multiparameter eigenvalue problems, *J. Math. Anal. Appl.* **92**(1983), 96–113.
21. P. Binding, Abstract oscillation theorems for multiparameter eigenvalue problems, *J. Differential Equations*, **49**(1983), 331-343.
22. P. Binding, Nonuniform right definiteness, *J. Math. Anal. Appl.* **102**(1984), 233-243.
23. P. Binding, Fundamental eigenvalues for one equation in several parameters, with application to linked equations, in *Multiparameter Problems*, Shiva Mathematics Series, Vol. 8, Nantwich, 1984, 1–8.
24. P. Binding, Multiparameter root vectors, *Proc. Edinburgh Math. Soc.* **32**(1989), 19–29.
25. P. Binding, A canonical form for self-adjoint pencils in Hilbert space, *J. Integral Equations and Operator Theory* **12**(1989), 324–342.
26. P. Binding and P.J. Browne, Comparison cones for multiparameter eigenvalue problems, *J. Math. Anal. Appl.* **77**(1980), 132–149.
27. P. Binding and P.J. Browne, Spectral properties of two-parameter eigenvalue problems, *Proc. Roy. Soc. Edinburgh* **89A**(1981), 157–173.
28. P. Binding and P.J. Browne, Multiparameter Sturm theory, *Proc. Roy. Soc. Edinburgh* **99A**(1984), 173–184.
29. P. Binding and P.J. Browne, Spectral properties of two-parameter eigenvalue problems II, *Proc. Roy. Soc. Edinburgh* **106A**(1987), 39–51.
30. P. Binding and P.J. Browne, Applications of two parameter-spectral theory to symmetric generalized eigenvalue problems, *Applicable Anal.* **29**(1988), 107–142.
31. P. Binding and P.J. Browne, Eigencurves for two-parameter self-adjoint ordinary differential equations of even order, *J. Differential Equations* **79**(1989), 289–303.
32. P. Binding and P.J. Browne, Asymptotics of eigencurves for second order ordinary differential equations I, II, preprints.
33. P. Binding, P.J. Browne, and L. Turyn, Existence conditions for two-parameter eigenvalue problems, *Proc. Roy. Soc. Edinburgh* **91A**(1981), 15–30.
34. P. Binding, P.J. Browne, and L. Turyn, Existence conditions for eigenvalue problems generated by compact multiparameter operators, *Proc. Roy. Soc. Edinburgh* **96A**(1984), 261–274.
35. P. Binding, P.J. Browne, and L. Turyn, Spectra properties of compact multiparameter operators, *Proc. Roy. Soc. Edinburgh* **98A**(1984), 291–303.
36. P. Binding, P.J. Browne, and L. Turyn, Existence conditions for higher order eigensets of multiparameter operators, *Proc. Roy. Soc. Edinburgh* **103A**(1986), 137–146.
37. P. Binding, A. Källström, and B.D. Sleeman, An abstract multiparameter spectral theory, *Proc. Roy. Soc. Edinburgh* **92A**(1982), 193–204.
38. P. Binding and K. Seddighi, Elliptic multiparameter eigenvalue problems, *Proc. Edinburgh Math. Soc.* **30**(1987), 215–228.
39. P. Binding and K. Seddighi, On root vectors of self-adjoint pencils, *J. Funct. Anal.* **70**(1987), 117–125.
40. P. Binding and H. Volkmer, Existence and uniqueness of indexed multiparametric eigenvalues, *J. Math. Anal. Appl.* **116**(1986), 131–146.
41. M.S. Birman and M.Z. Solomjak, Spectral asymptotics of nonsmooth elliptic operators. II, *Trans. Moscow Math. Soc.* **28**(1973), 1–32.
42. M.S. Birman and M.Z. Solomjak, Asymptotic behaviour of the spectrum of differential equations, *J. Soviet Math.* **12**(1979), 247–282.

43. M.S. Birman and M.Z. Solomjak, *Quantitative analysis in Sobolev imbedding theorems and applications to spectral theory*, Amer. Math. Soc. Transl. (2) **114**(1980).
44. M. Bôcher, The theorems of oscillation of Sturm and Klein. I, *Bull. Amer. Math. Soc.* **4**(1897/98), 295–313.
45. M. Bôcher, The theorems of oscillation of Sturm and Klein. II, *Bull. Amer. Math. Soc.* **4**(1897/98), 365–376.
46. M. Bôcher, The theorems of oscillation of Sturm and Klein. III, *Bull. Amer. Math. Soc.* **5**(1897/98), 22–43.
47. J. Bognár, *Indefinite Inner Product Spaces*, Springer, New York, 1974.
48. K. Borsuk. Drei sätze über die n-dimensionale euklidische sphäre, *Fund. Math.* **20**(1933), 177–190.
49. P.J. Browne, A multi-parameter-eigenvalue problem, *J. Math. Anal. Appl.* **38**(1972), 553–568.
50. C.C. Camp, A expansion involving P inseperable parameters associated with a partial differential equation, *Amer. J. Math.* **50**(1928), 259–268.
51. C.C. Camp, On multiparameter expansions associated with a differential system and auxiliary conditions at several points in each variable, *Amer. J. Math.* **60**(1930), 447–452.
52. R.D. Carmichael, Boundary value and expansion problems: algebraic basis of the theory, *Amer. J. Math.* **43**(1921), 69–101.
53. R.D. Carmichael, Boundary value and expansion problems: formulation of various transcendental problems, *Amer. J. Math.* **43**(1921), 232–270.
54. R.D. Carmichael, Boundary value and expansion problems: oscillatory, comparison, and expansion problems, *Amer. J. Math.* **44**(1922), 129-152.
55. E.A. Coddington and N. Levinson, *Theory of ordinary diffential equations*, McGraw-Hill, New York, 1955.
56. H.O. Cordes, Separation der Variablen in Hilbertschen Räumen, *Math. Ann.* **125**(1953), 401–434.
57. H.O. Cordes, Über die Spektralzerlegung von hypermaximalen Operatoren, die durch Separation der Variablen zerfallen. I, *Math. Ann.* **128**(1954/55), 257–289.
58. H.O. Cordes, Über die Spektralzerlegung von hypermaximalen Operatoren, die durch Separation der Variablen zerfallen. II, *Math. Ann.* **128**(1954/55), 373–411.
59. B. Ćurgus and H. Langer, A Krein space approach to symmetric ordinary differential operators with an indefinite weight function, *J. Differential Equations* **79**(1989), 31–61.
60. K. Daho and H. Langer, Sturm-Liouville operators with an indefinite weight function, *Proc. Roy. Soc. Edinburgh* **78A**(1977), 161–191.
61. A.C. Dixon, Harmonic expansions of functions of two variables, *Proc. London Math. Soc.* **5**(1907), 411–478.
62. H.P. Doole, A certain multiparameter expansion, *Bull. Amer. Math. Soc.* **37**(1931), 439–446.
63. N. Dunford and J.T. Schwartz, *Linear Operators*, Part II, Wiley, New York, 1963.
64. W. Eberhard, G. Freiling, and A. Schneider, On the distribution of the eigenvalues of a class of indefinite eigenvalue problems, *Differential Integral Equations* **3**(1990), 1167–1179.
65. W. Eberhard, G. Freiling, and A. Schneider, Expansion theorems for a class of regular indefinite eigenvalue problems, *Differential Integral Equations* **3**(1990), 1181–1200.

66. D.E. Edmunds and W.D. Evans, *Spectral Theory and Differential Operators*, Clarendon Press, Oxford, 1987.
67. W.N. Everitt, M.K. Kwong, and A. Zettl, Oscillation of eigenfunctions of weighted regular Sturm-Liouville problems, *J. London Math. Soc.* **27**(1983), 106–120.
68. W.H. Everitt, M.K. Kwong, and A. Zettl, Differential operators and quadratic inequalities with a degenerate weight, *J. Math. Anal. Appl.* **98**(1984), 378–399.
69. M. Faierman, The completeness and expansion theorem associated with the multiparameter eigenvalue problem in ordinary differential equations, *J. Differential Equations* **5**(1969), 197–213.
70. M. Faierman, Asymptotic formulae for the eigenvalues of a two-parameter ordinary differential equation of the second order, *Trans. Amer. Math. Soc.* **168**(1972), 1–52.
71. M. Faierman, Asymptotic formulae for the eigenvalues for a two-parameter system of ordinary differential equations of the second order, *Canad. Math. Bull.* **17**(1975), 657–665.
72. M. Faierman, On the distribution of the eigenvalues of a two-parameter system of ordinary differential equations of the second order, *SIAM J. Math. Anal.* **8**(1977), 854–870.
73. M. Faierman, Eigenfunction expansions associated with a two-parameter system of differential equations, *Proc. Roy. Soc. Edinburgh* **81A**(1978), 79–93.
74. M. Faierman, An oscillation theorem for a two-parameter system of differential equations, *Quaestiones Math.* **3**(1979), 313–321.
75. M. Faierman, An eigenfunction expansion associated with a two-parameter system of differential equations. I, *Proc. Roy. Soc. Edinburgh* **89A**(1981), 143–155.
76. M. Faierman, An eigenfunction expansion associated with a two-parameter system of differential equations, in *Spectral Theory of Differential Operators*, Math. Studies, Vol. 55, North-Holland, Amsterdam, 1981, 169–172.
77. M. Faierman, An eigenfunction expansion associated with a two-parameter system of differential equations. II, *Proc. Roy. Soc. Edinburgh* **92A**(1982), 87–93.
78. M. Faierman, An eigenfunction expansion associated with a two-parameter system of differential equations. III, *Proc. Roy. Soc. Edinburgh* **93A**(1983), 189–195.
79. M. Faierman, A left definite two-parameter eigenvalue problem, in *Differential Equations*, Math. Studies, Vol. 92, North-Holland, Amsterdam, 1984, 205–211.
80. M. Faierman, The expansion theorem for a left definite two-parameter system of ordinary differential equations of the second order, in *Multiparameter Problems*, Shiva Mathematics Series, Vol. 8, Nantwich, 1984, 13–20.
81. M. Faierman, The eigenvalues of a multiparameter system of differential equations, *Applicable Anal.* **19**(1985), 275–290.
82. M. Faierman, Expansions in eigenfunctions of a two-parameter system of differential equations, *Quaestiones Math.* **10**(1986), 135–152.
83. M. Faierman, Expansions in eigenfunctions of a two-parameter system of differential equations. II, *Quaestiones Math.* **10**(1987), 217–249.
84. M. Faierman, Regularity of solutions of an elliptic boundary value problem in a rectangle, *Comm. Partial Differential Equations* **12**(1987), 285–305.
85. M. Faierman, Elliptic problems involving an indefinite weight, *Trans. Amer. Math. Soc.* **320**(1990), 253–279.
86. M. Faierman, A two-parameter eigenvalue problem involving complex potentials, *Proc. Roy. Soc. Edinburgh* **116A**(1990), 177-191.

87. M. Faierman, On the principal subspaces associated with a two-parameter eigenvalue problem, in Boundary Value Problems for Ordinary Differential Equations (to appear).
88. M. Faierman and G.F. Roach, Linear elliptic eigenvalue problems involving an indefinite weight, *J. Math. Anal. Appl.* **126**(1987), 517–528.
89. M. Faierman and G.F. Roach, Full and partial-range eigenfunction expansions for a multiparameter system of differential equations, *Applicable Anal.* **28**(1988), 15–37.
90. M. Faierman and G.F. Roach, Eigenfunction expansions for a two-parameter system of differential equations, *Quaestiones Math.* **12**(1988), 65–99.
91. M. Faierman and G.F. Roach, Full and half-range eigenfunction expansions for an elliptic boundary value problem involving an indefinite weight, in *Differential Equations*, Lecture Notes in Pure and Appl. Math., Vol. 118, Dekker, New York, 1989, 231–236.
92. M. Faierman and G.F. Roach, Eigenfunction expansions associated with a multiparameter system of differential equations, *Differential Integral Equations* **2**(1989), 45–56.
93. J. Fleckinger-Pelle, Asymptotics of eigenvalues for some non-definite elliptic problems, in *Ordinary and Partial Differential Equations*, Lecture Notes in Math., Vol. 1151, Springer, New York, 1985, 148–156.
94. J. Fleckinger and M.L. Lapidus, Eigenvalues of elliptic boundary value problems with an indefinite weight function, *Trans. Amer. Math. Soc.* **295**(1986), 305–324.
95. J. Fleckinger and A.B. Mingarelli, On the eigenvalues of non-definite elliptic operators, in *Differential Equations*, Math. Studies, Vol. 92, North-Holland, Amsterdam, 1984, 219–227.
96. G. Freiling and F.J. Kaufmann, *On uniform and L^p-convergence of eigenfunction expansions for indefinite eigenvalue problems*, Schriftenreihe des fachbereichs Mathematik 150, Universität Duisburg Gesamthochschule, 1989.
97. G.A. Gadzhiev, On a multitime equation and its reduction to a multiparameter spectral problem, *Soviet Math. Dokl.* **32**(1985), 710–713.
98. I.C. Gohberg and M.G. Krein, *Introduction to the Theory of Linear Nonselfadjoint Operators*, Amer. Math. Soc., Providence, R.I., 1969.
99. I.C. Gohberg and M.G. Krein, *Theory and Applications of Volterra Operators in Hilbert Space*, Amer. Math. Soc., Providence, R.I., 1970.
100. S. Goldberg, *Unbounded Linear Operators*, McGraw-Hill, New York, 1966.
101. P. Grisvard, *Elliptic Problems in Nonsmooth Domains*, Pitman, London, 1985.
102. O. Haupt, *Untersuchungen über Oszillationstheoreme*, Teubner, Leipzig, 1911.
103. O. Haupt, Über eine Methode zum Bewise von Oszillationstheoreme, *Math. Ann.* **76**(1915), 67–104.
104. P. Hess, On the relative completeness of the generalized eigenvectors of elliptic eigenvalue problems with indefinite weight functions, *Math. Ann.* **270**(1985), 467–475.
105. E. Hilb, Eine Erweiterung des Kleinschen Oszillationstheorems, *Jahresbericht d.d. Math. Ver.* **16**(1907), 279–285.
106. D. Hilbert, *Grundzüge einer Allgemeiner Theorie der Linearen Integralgleichungen*, Chelsea, New York, 1953.
107. E. Hille, *Lectures on Ordinary Differential Equations*, Addison-Wesley, Reading, Mass., 1969.

108. E. Hölmgren, Über Randwertaufgaben bei einer linearen Differentialgleichungen zweiter Ordnung, *Ark. Mat., Astro. och Fysik* **1**(1904), 401–417.
109. L. Hörmander, Uniqueness theorems for second order elliptic differential equations, *Comm. Partial Equations* **8**(1983), 21–64.
110. I.S. Iohvidov, On the spectra of Hermitian and unitary operators in a space with an indefinite metric, *Dokl. Akad. Nauk SSSR* **71**(1950), 225–228.
111. I.S. Iohvidov, M.G. Krein, and H. Langer, *Introduction to the Spectral Theory of Operators in Spaces with an Indefinite Metric*, Akademie, Berlin, 1982.
112. G.A. Isaev, On root elements of multiparameter spectral problems, *Soviet Math. Dokl.***21** (1980), 127–130.
113. F. John, *Plane Waves and Spherical Means Applied to Partial Differential Equations*, Interscience, New York, 1955.
114. K. Jörgens, *Spectral Theory of Second Order Ordinary Differential Equations*, Aarhus Universitet Lecture Notes, 1964.
115. J. Kadlec, On the regularity of the solution of the Poisson problem on a region with boundary similar to the boundary of a cube, *Czechoslovak Math. J.* **13**(1963), 591–611.
116. A. Källström and B.D. Sleeman, A multi-parameter Sturm-Liouville problem, in *Ordinary and Partial Differential Equations*, Lecture notes in Math., Vol. 415, Springer, New York, 1974, 394–401.
117. A. Källström and B.D. Sleeman, A left definite multiparameter eigenvalue problem in ordinary differential equations, *Proc. Roy. Soc. Edinburgh* **74A**(1974/75), 145–155.
118. H.G. Kaper, M.K. Kwong, C.G. Lekkerkerker, and A. Zettl, Full- and partial range eigenfunction expansions for Sturm-Liouville problems with indefinite weights, *Proc. Roy. Soc. Edinburgh* **98A**(1984), 69–88.
119. H.G. Kaper, M.K. Kwong, and A. Zettl, Singular Sturm-Liouville problems with nonnegative and indefinite weights, *Monat. Math.* **97**(1984), 177–189.
120. T. Kato, *Perturbation Theory for Linear Operators*, 2nd edn., Springer, New York, 1976.
121. F. Klein, Über Körper, welche von confocalen Flächen zweiten Grades begrenzt sind, *Math. Ann.* **18**, (1881), 410–427.
122. F. Klein, *Gesammelte Mathematische Abhandlungen*, Vol. 2, Springer, Berlin, 1922.
123. V.A. Kondrat'ev and O.A. Oleinik, Boundary-value problems for partial differential equations in non-smooth domains, *Russian Math. Surveys* **38**(1983), 1–86.
124. O.A. Ladyzhenskaya and N.N. Ural'tseva, *Linear and Quasilinear Elliptic Equations*, Academic, New York, 1968.
125. A.I. Mal'cev, *Foundations of Linear Algebra*, Freeman, San Francisco, 1963.
126. A.S. Markus, *Introduction to the Spectral Theory of Polynomial Operator Pencils*, Amer. Math. Soc., Providence, R.I., 1988.
127. V.G. Maz'ja, *Sobolev Spaces*, Springer, New York, 1985.
128. A.B. Mingarelli, Indefinite Sturm-Liouville problems, in *Ordinary and Partial Differential Equations*, Lecture Notes in Math., Vol. 964, Springer, New York, 1983, 519–528.
129. A.B. Mingarelli, *Volterra-Stieltjes Integral Equations and Generalized Ordinary Differential Expressions*, Lecture Notes in Math., Vol. 989, Springer, New York, 1983.
130. A.B. Mingarelli, On the existence of nonsimple real eigenvalues for general Sturm-Liouville problems, *Proc. Amer. Math. Soc.* **89**(1983), 457–460.

131. A.B. Mingarelli, Asymptotic distribution of the eigenvalues of non-definite Sturm-Liouville problems, in *Ordinary and Partial Differential Equations*, Lecture Notes in Math., Vol. 1032, Springer, New York, 1983, 375–383.
132. A.B. Mingarelli, The non-real point spectrum of generalized eigenvalue problems, *C.R. Math. Rep. Acad. Sci. Canada* **6**(1984), 117–121.
133. A.B. Mingarelli, A survey of the regular weighted Sturm-Liouville problem — the non-definite case, in *Proc. of the Workshop on Applied Differential Equations*, World Scientific Publishing, Singapore and Philadelphia, 1986, 109–137.
134. S. Mizohata, *The Theory of Partial Differential Equations*, University Press, Cambridge, 1973.
135. M.A. Naimark, *Linear Differential Operators*, Part I, Ungar, New York, 1967.
136. M.A. Naimark, *Linear Differential Operators*, Part II, Ungar, New York, 1968.,
137. A.J. Pell, Linear equations with two parameters, *Trans. Amer. Math. Soc.* **23**(1922), 198–211.
138. Å. Pleijel, Sur la distribution des valeurs propres de problèmes régis par l'équation $\Delta u + \lambda k(x,y)u = 0$, *Ark. Mat., Astr. och Fysik* **29B**(1942), 1–8.
139. Å. Pleijel, Le problème spectral de certaines équations aux dérivées partielles, *Ark. Mat., Astr. och Fysik* **30A**(1944), 1–47.
140. R.G.D. Richardson, Theorems of oscillation for two linear differential equations of the second order with two parameters, *Trans. Amer. Math. Soc.* **13**(1912), 22–34.
141. R.G.D. Richardson, Über die notwendigen und hindreichenden Bedingungen für das Bestehen eines Kleinschen Oszillationstheorms, *Math. Ann.* **73**(1913), 289–304. Erratum:ibid **74**, 312.
142. R.G.D. Richardson, Contributions to the study of oscillation properties of the solutions of linear differential equations of the second order, *Amer. J. Math.* **40**(1918), 283–316.
143. B.D. Sleeman, Completeness and expansion theorems for a two-parameter eigenvalue problem in ordinary differential equations using variational principles, *J. London Math. Soc.* **6**(1973), 705–712.
144. B.D. Sleeman, Left definite multiparameter eigenvalue problems, in *Spectral Theory and Differential Equations*, Lecture Notes in Math., Vol. 448, Springer, New York, 1975, 307–321.
145. B.D. Sleeman, Multiparameter spectral theory in Hilbert space, *J. Math. Anal. Appl.* **65**(1978), 511–530.
146. B.D. Sleeman, *Multiparameter Spectral Theory in Hilbert Space*, Pitman, London, 1978.
147. B.D. Sleeman, Klein oscillation theorems for multiparameter eigenvalue problems in ordinary differential equations, *Nieuw Arch. Wisk.* **27**(1979), 341-362.
148. C. Sturm, Sur les équations différentielles du second ordre, *J. de Math.* **1**(1836), 106-186.
149. E.C. Titchmarsh, *Eigenfunction Expansions*, Part I, 2nd edn., Clarendon, Oxford, 1962.
150. L. Turyn, Sturm-Liouville problems with several parameters, *J. Differential Equations* **38**(1980), 239–259.
151. H. Volkmer, On the completeness of eigenvectors of right definite multiparameter problems, *Proc. Roy. Soc. Edinburgh* **96A**(1984), 69–78.
152. H. Volkmer, *Multiparameter Eigenvalue Problems and Expansion Theorems*, Lecture Notes in Math., Vol. 1356, Springer, New York, 1988.

153. H.F. Weinberger, *Variational Methods for Eigenvalue Approximation*, CBMS Regional Conf. Ser. Appl. Math., Vol. 15, SIAM, Philadelphia, 1974.
154. A.C. Zaanen, *Linear Analysis*, North-Holland, Amsterdam, 1953.